Statistics for Geoscientists

Statistics for Geoscientists

by

DIETER MARSAL

University of Stuttgart
Federal Republic of Germany

Translation editor

DANIEL F. MERRIAM

Wichita State University, USA

PERGAMON PRESS

OXFORD · NEW YORK · BEIJING · FRANKFURT
SÃO PAULO · SYDNEY · TOKYO · TORONTO

U.K.	Pergamon Press, Headington Hill Hall, Oxford OX3 0BW, England
U.S.A.	Pergamon Press, Maxwell House, Fairview Park, Elmsford, New York 10523, U.S.A.
PEOPLE'S REPUBLIC OF CHINA	Pergamon Press, Room 4037, Qianmen Hotel, Beijing, People's Republic of China
FEDERAL REPUBLIC OF GERMANY	Pergamon Press, Hammerweg 6, D-6242 Kronberg, Federal Republic of Germany
BRAZIL	Pergamon Editora, Rua Eça de Queiros, 346, CEP 04011, Paraiso, São Paulo, Brazil
AUSTRALIA	Pergamon Press Australia, P.O. Box 544, Potts Point, N.S.W. 2011, Australia
JAPAN	Pergamon Press, 8th Floor, Matsuoka Central Building, 1-7-1 Nishishinjuku, Shinjuku-ku, Tokyo 160, Japan
CANADA	Pergamon Press Canada, Suite No. 271, 253 College Street, Toronto, Ontario, Canada M5T 1R5

Copyright © 1987 Pergamon Books Ltd.

First English edition 1987

This book is a translation of *Statistische Methoden für Erdwissenschaftler* 1979, published by E. Schweizerbart'sche Verlagsbuchhaudlung (Nägele u. Obermiller) Stuttgart. ©1967 by E. Schweizerbart'sche Verlagsbuchhaudlung (Nägele u. Obermiller), Stuttgart.

Library of Congress Cataloging-in-Publication Data

Marsal, Dieter
Statistics for geoscientists.
Translation of: Statistische Methoden für Erdwissenschaftler.
Bibliography: p.
Includes indexes.
1. Earth sciences—Statistical methods.
I. Merriam, Daniel Francis. II. Title.
QE33.2.M3M3413   1987   550'.72   87–10582

British Library Cataloguing in Publication Data

Marsal, D.
Statistics for geoscientists.
1. Earth sciences—Statistical methods
I. Title II. II. Statistische Methoden für Erdwissenschaftler.
English
519.5'02455   QE33.2.M3

ISBN 0–08–026268–6 (Hardcover)
ISBN 0–08–026260–0 (Flexicover)

Reproduced, printed and bound in Great Britain by
Hazell Watson & Viney Limited,
Member of the BPCC Group,
Aylesbury, Bucks

To my dear old friend
Dr. Karl Dietrich Adam
Professor of Paleontology and Prehistory

Foreword to the English Edition

The use of statistics is becoming more evident in geology and related earth sciences. Witness the many texts and references on the subject as well as the research papers and special meetings such as the one at the Geological Society of America in Reno on "The Use and Abuse of Statistics." Because of the importance in understanding the background of such things as sampling, distributions, significance, correlation, classification, etc., geologists need to be introduced to the subject early in their careers. This little book is intended for just that purpose.

Simple, straightforward, and concise, this book on *Statistics for Geoscientists* by Dieter Marsal is ideal as an introduction to the subject. The first edition was published in 1967 in German and it was revised in 1979. The elegance of the presentation, however, was lost on the English-speaking geological public and the book's influence was limited. Although there are many examples, the book is not a step-by-step cookbook on how to use statistics. Subject matter covers everything from sampling, distributions, and central tendencies through significance tests, time series, Markov chains, correlation and regression, and Fourier analysis to discriminant analysis, splines, and the analysis of variance. A brief but thorough presentation, it is sprinkled liberally with examples of application, albeit European ones.

For some readers, the compact style (23 chapters) may be difficult to adjust to, and for others the lack of familiar examples may be distracting, but for most, the directness, completeness, and brevity will be welcome. The book can be used as a first approach to a difficult subject, for clarification of particular subjects, or for new ideas in applications. Some may ask why another statistics book, especially one first published so long ago? The answer is easy—people learn from different approaches and this approach is different from others that are available. Although the book was published first 20 years ago, the basic tenets and mathematics have not changed, only the adaptations to the computer which have improved during this time, especially in the presentation of output. So, it is believed the book has a place and will be appreciated, even though there are other books on the subject.

I first encountered Marsal's book soon after its original publication, and although the German was not difficult, I thought how much better it would be if translated into English. We worked toward that goal—an English edition—for several years. Finally, Dieter did the rewrite himself with the stipulation I edit the English version; I gladly agreed. Those familiar with the German edition will note the English edition is not a word-for-word translation but subject-by-subject translation. During the project it was realized that some changes were needed, and those were affected. I would like to thank my German-speaking colleague, Dr. Peter G. Sutterlin of Wichita State University, for reading the manuscript and making many helpful suggestions. His thoughtful contribution to the project was most appreciated.

So, here is yet another statistics book for earth scientists—from a European viewpoint. May geologists learn to appreciate the importance of statistics and their application in the earth sciences!

Department of Geology,   D. F. MERRIAM
Wichita State University,
Wichita, Kansas 67208, USA

Contents

CHAPTER 1

Introduction, Scope and Purpose of Applied Statistics

Applied statistics are concerned mainly with describing and analyzing the *variability* of various data gathered from a larger collection termed a *universe* or *population*. For instance, the universe might be defined as a sand dune composed of grains of various shapes and sizes that we want to study by scrutinizing a finite number of grains of a gathered *sample*.

Invariably, every statistical investigation commences with the collection of raw data and its representation in a clearly arranged mode. The raw data may have been obtained by collecting observations or measurements; they may refer to nonnumerical qualities such as the colors and morphological properties of a mineral, or they may comprise numerical quantities such as lengths and weights. All data must be stored in a *databank* or catalogued as a *list*. This *inventory* is a document representing a basic set of data from which different types of inferences may be derived by analysts.

In most situations—especially in elementary statistics—the first step of analysis consists in determining the *frequency distribution* of the collected data. It shows the frequency of occurrence of each listed value or quality, or it shows the frequency of occurrence of all listed numbers which are in a certain *interval*, also termed *a class* or *fraction*.

Many empirical frequency distributions can be described more or less accurately by functions with well-known mathematical properties. Typical examples are the *normal distribution* which seems to be the most usual mathematical distribution function, the *binomial distribution* which describes the expected frequencies of the occurrence of two alternative states, and the *distribution of Poisson* which can be shown to describe quantitatively the occurrence of rare events.

Not much can be learned directly from the initial raw-data list; however, *condensed* and systematically arranged as a frequency distribution, the variability of the considered property or properties can be seen at a glance, especially if plotted as a *histogram*. Yet it may be desirable to condense the data further. This is done by characterizing the frequency distribution by a few numbers termed *parameters* which describe properties of the distribution as a whole. Most important are average values such as the *mean* and *median*, and variability indicators such as the *variance* and *standard deviation*. These and other concepts related to them will be discussed at the appropriate places. Some of these parameters characterizing a sampling distribution especially are appropriate to the geological sciences.

To obtain additional results, the elementary methods of routine collection, tabulation, description, and condensation of data *per se*, are to be complemented by more advanced techniques.

In most situations, the number of individuals in a universe is large, and most parts of it are inaccessible. Thus, the investigator must be content with a sample whose size is small compared to the total number of individuals of the whole population. Unfortunately, the frequency distribution usually differs for each sample, and none is exactly alike to the frequency distribution of the sample universe. Hence, we cannot readily assume that the results derived from a sample are *representative* for its universe. This problem gives rise to the following questions:

1. How many samples must be drawn from a population to obtain representative results?
2. How large is the size of a representative sample? Strictly speaking: How many individuals constitute a representative sample?
3. If the objects of the universe are distributed spatially such as minerals in a rock or stars in a stellar cloud, then at what locations must the samples be collected?

In mathematical statistics these questions are dealt with on the assumption that the objects of the universe are *randomly distributed*, that is distributed according to the laws of chance. Then the problems can be solved by applying the principles of probability theory to statistics. This part of mathematical statistics is termed *sampling theory*. It turns out that the confidence we can have in the results inferred from a sample depends on the size of the sample and the universe's frequency distribution. Unfortunately, however, probability theory never gives answers that are certain; the answers are invariably of the type "With a specified certainty of less than 100%, the mean or the variance, etc., of the universe is contained in some interval, the so-called *confidence interval.*" For instance, the answer may be: "With a certainty of 99%, the mean grain size of an investigated dune is between 0.11 and 0.19 centimetres." Thus the expression "with a certainty of 99%" indicates that the *estimate* of the confidence interval is correct in 99 situations out of 100 drawn samples*. In the latter situation, the mean grain size of the universe is either larger than 0.19 cm or smaller than 0.11 cm.

---

*The colloquial expression "with a certainty of 99%" corresponds to the technical jargon "for the *confidence coefficient* 0.99."

The probabilistic approach to sampling poses two fundamental questions:

1. What is the appropriate confidence coefficient? 0.9, or 0.95, or 0.99, or 0.999, etc.?
2. Is the basic assumption of sampling theory—randomness of the distribution—justified?

When attempting to forecast the results of a democratic election by polling a sample of people, the appropriate confidence coefficient may be determined by comparing forecasts with corresponding outcomes. This approach, however, is seldom workable in the geosciences. Thus, there is no unique answer to the first question—the investigator has to rely on experience, and the answer may be different for different geological universes. Hence, statistics as applied to the earth sciences is—although mainly a rigorous discipline—partly an art.

The answer to the second question is Yes and No. Occasionally, a geological body is rather homogeneous, and randomness is a fair assumption. In most situations, however, geological bodies are rather inhomogeneous as a whole. The mineral distributions of sediments, magmatic and metamorphic rocks furnish an inexhaustable multitude of examples and realized possibilities ranging from ideal randomness to almost regular pattern arrangements.

Another important aspect of sampling theory is the *testing of statistical hypotheses*. In the simplest situation, two different distributions are compared with each other to test whether they belong to the same universe. This is done usually by applying the *Chi-square method*. In many instances, however, it is sufficient to compare the means and variances by applying the *Student-t test* and *Snedecor's F-test* respectively or some other technique*. Qualitative frequency distributions, where the frequency is characterized by designations as "abundant," "rare," etc., are compared with each other by employing *ranking methods*. The idea behind the ranking is to allocate a placement to each frequency designation in much the same manner as in school the "1" or "A" might be assigned to note "excellent", "2" or "B" good, etc.

These methods may fail and thus the question whether two samples belong to the same population remains undecided. Then the technique of *discriminant analysis* might be applied successfully. It is based on the idea that a proper combination of several properties may be more informative than a comparison of a single property.

The advanced techniques of applied statistics permit the simultaneous

---

*The Student-t test is based on a mathematical distribution function that was published under the pseudonym "Student." This piece of work can be considered the starting point of modern statistics.

comparison of an arbitrary number of samples. Thereby the universe is tested for homogeneity, and the population's variance is split and allocated to different determining factors. This permits in the determination of the underlying factors which mainly are responsible for the variablity of the universe. Typical keywords of this powerful technique are *"analysis of variance," "factorial analysis,"* and *"Latin squares."* These and related topics are covered amply by an extensive literature and their understanding requires mathematical maturity. Hence, in an introductory text only the simpler methods can be discussed. The modern rather sophisticated approaches to factorial analysis are beyond the scope of this book.

It may be that several properties of a universe are interrelated as possibly, for instance, the largest and the smallest diameter of a mineral or the size of some fossil and geological time. Usually the interrelationship is obscured by unknown influences. As a consequence, the relationship among the properties under consideration cannot be described exactly but only grossly by simple graphs such as straight lines, tilted planes, parabolas, exponential curves etc. This relationship is known as *"correlation."* The concept has two main aspects, a statistical and a numerical one. For instance, when plotting the diameters of *Cosmoceras* individuals versus geological time, we may obtain a set of points scattered about a straight line. Then the first task is to determine a *"regression line"* such that it represents a *best fit* of the data. Now the solution to the question "What is the best fit and how can the best fitting curve be obtained?" is determined by applying elementary numerical analysis methods. In most situations, either the *Gaussian least-square method* or the modern approach of employing *spline functions* is used.

The statistical aspect of this type of correlation is concerned with the interpretation of interrelated data. How good is the fit? Are the regression curves of a sample representative for the universe? How are the regression curves related to the frequency distribution which encompasses all correlated properties of the population?

Obviously, these questions are rather technical. The reader might be more interested in detecting causal relationships among correlated data—a topic popular in the medical sciences. Primarily, a correlation constitutes a formal measure which might suggest, but does not prove necessarily a cause-and-effect relationship. Thus the statements "The Dutch coast is sinking slowly every year" and "The population of the Netherlands is steadily increasing" certainly may constitute a correlation, but is not necessarily a causal relationship. The statistician always must be on guard when attempting to draw conclusions which are beyond the scope of statistical inference. Physico-biological interpretations of statistical results must include supporting data or complementary reasoning.

Variability of various properties is one of the outstanding features of most geological bodies. Hence statistics as the science of studying the variability of observed data should be a major tool of the geoscientist. However, statistical inference is not a method in itself and although perhaps necessary, usually is not sufficient from which to draw causal conclusions.

In a *tour de force* I have mentioned most topics dealt with in this book with the exception of such topics as *analysis of time series* and *banking cycles*. Advanced methods and special topics such as the analysis of random vibrations as applied to geophysics are not included. The inclusion of these would have made the book bulky and too lengthy and therefore unattractive to the beginner.

Further there is no introduction to probability theory. My reason is simple. As soon as this mathematical discipline is treated beyond the theory of tossing coins, it becomes intricate and demanding. To understand, for instance, probabilistic sampling theory requires a mathematical maturity which cannot be expected from the practicing geologist, petrographer, or paleontologist. Thus, the reader will become better acquainted with the application of statistical techniques rather than with the derivation of the equations developed from basic principles.

CHAPTER 2

Classification and Tabulation of Frequency Distributions

For a collection of data, the number of items in a given class C is termed C's *frequency*. In the geosciences, C may be a nonnumerical property such as a color, a specific type of object such as a mineral, a single numerical value, or an interval such as the set of all numbers between zero and one. The frequency may refer to a *universe* as a whole or to a *sample*. The hierarchical order of most statistical properties is covered by the following scheme:

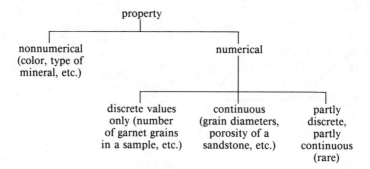

When a collection of data is separated into several classes, the number of items in a given class is the *absolute frequency*, and the absolute frequency divided by the total number of items is the *relative frequency*. The sum of all relative frequencies is unity (100%). The set of all frequencies with their corresponding classes is the *absolute (relative) frequency distribution*. Occasionally, the exact frequencies are unknown but can be designated qualitatively by expressions such as "rare" and "abundant."

In most situations, the investigator is interested in relative frequencies rather than in absolute ones. However, the total numbers of items must be known to estimate whether the sample distribution is representative of the universe. Thus, whenever possible, the total number of items involved should be stated.

Example 2.1   Distributions with semiquantitative or nonnumerical frequencies

Remains of some land mammals which are present in the Devil's Hole (Teufelslucken) near Eggenburg in Lower Austria (Adam, 1966):

Remains	Cave-hyena	Mammoth
Bones	1096	7
Teeth	955	31

This is a distribution with four classes (cave-hyena bones, etc.) and a total of 2089 items. By a rather complex analysis of the recovered material, the minimum number of animals that were dragged into or perished in the cave could be determined:

cave-hyena: at least 67   mammoth: at least 13.

This is a frequency distribution for two classes (a so-called *dichotomous distribution*). In the following example, frequencies of occurrence of mammoth remains are indicated by the signs + (a few items) and − (none):

Czechoslovakia, Magdalénien

Location	Remains of mammoth
Adlerova jeskyně	−
Balcarova skála	+
Hadi jeskyně	−
....	.....
Žitného jeskyně	−

This semiquantitative distribution has an interesting interpretation: At the end of the Ice Ages the mammoth had ceased to be a factor in hunting for man. The remains of mammoth at the Devil's Hole near Eggenburg are evidence of hunting at an earlier time.

Example 2.2   Distributions with nonnumerical classes

The left-hand distribution refers to the relative abundancy of minerals in a thin section that were estimated by comparison with a chart for visual percentage evaluation (cf. Müller, 1964, p. 143; Niggli, 1948, p. 149).

Green schist from Furulund,
Norway (Eskola, 1939, p. 346)

Mineral	Relative frequency, %
Quartz	1.1
Albite	39.9
Chlorite	29.4
Epidote	23.0
Hornblende	3.5
Calcite, etc.	2.6
Sum	99.5%

Note: Absolute frequencies are not defined; the accuracy of the relative frequencies do not seem to be justified.

Mammoth molars from the
Devil's Hole (Adam, 1966)

Molar	Frequency, count
ml	3
m2	4
m3	7
M1	1
M2	1
M3	1
Sum	17

m = milk dentition;
M = molars of adult individuals.

Example 2.3   A bivariate distribution

An *univariate distribution* depends on one variable, a *bivariate distribution* on two. The frequency distributions of the last example are univariate: the variables are the type of mineral and type of teeth respectively. The first distribution of Example 2.1 is bivariate: the first variable describes different types of remains, the second refers to different types of animals. The following distribution depends on one numerical variable and a nonnumerical one.

Main minerals of a glauconitic sediment in % per volume (Correns, 1939, p. 209)

Fraction (μ)	2–11	11–20	20–110	110–200
Quartz	15	<25	55	>60
Feldspar	10	<15	<20	10
Mica	> 5	5	< 3	0
Ore	< 3	< 3	< 3	0
Glauconite	<50	>45	<20	<20

Note: Absolute frequencies are not defined.

Example 2.4   The tabulation of discrete frequency distributions

At twenty different locations designated by letters, samples were collected from a sandstone. After extracting the heavy minerals from a fine-sand fraction, the number of garnets per 100 grains were counted.

a:1	b:6	c:0	d:2	e:2	f:1	g:3	h:1	i:3	j:2
k:7	l:1	m:0	n:2	o:2	p:4	q:2	r:3	s:2	t:1

In the following tables, a dash represents one location, the sign ⧾⧾⧾ five locations.

Locations: all

Number of garnets	Frequency	
0	//	2
1	⧾⧾⧾	5
2	⧾⧾⧾ //	7
3	///	3
4	/	1
5		0
6	/	1
7	/	1

Locations: a, d, e, f, g, h, l n, p, s, t

Number of garnets	Frequency	
0		0
1	⧾⧾⧾	5
2	///	3
3	//	2
4	/	1
5		0
6		0
7		0

The example shows that a frequency distribution might be affected by the selection and number of locations. The left-hand distribution is more spread out than the right one, and the positions of the two maxima differ. Both distributions are termed *discrete* because the variable—the number of garnets—can only assume discrete values.

Example 2.5   The tabulation of a continuous frequency distribution

The following list shows porosity values measured in sandstone cores from a depth ranging between 1100 and 1105 meters (Engelhardt, 1960, fig. 11)*:

22.1	23.5	25.3	26.6	23.9	26.0	22.8	22.3	23.1	23.0	21.0	21.8
22.0	22.2	22.3	22.4	22.4	22.4	22.3	21.6	22.1	22.6	22.1	21.9
22.3	23.9	23.2	22.5	23.7	23.3	24.4	22.6	23.9	24.2	27.6	27.9
25.2	21.7	20.0	19.8	21.5	25.6	25.3	24.1	28.6	23.7	24.0	21.8
24.9	24.2	25.0	23.7	27.3	23.0	23.8	21.2	21.1%			

In principle at least, the porosity of a rock or any other porous medium can assume any value between 0 and 1. It therefore would make no sense to represent the sample by a discrete distribution; rather we should organize the values by grouping them into classes. First *classes of equal width* will be used by selecting intervals of the length of 1%. To prepare the entries, we read the list of analyses and draw a dash for each value. If a porosity value corresponds to an endpoint of an interval, it will be included in the lower group. For instance, a porosity of 26% will be

*Porosity is defined as the ratio (pore volume)/(bulk volume).

assigned to the class 25–26% rather than to the class 26–27%. It goes
without saying that other conventions also may be employed to allocate
endpoint values to classes.

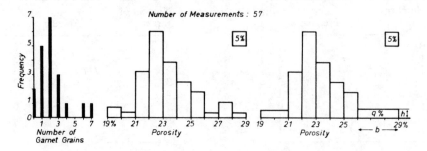

FIGURE 1. *Left*: Representation of discrete distribution. Frequencies are
plotted on ordinate. Diagram shown refers to left table of Example 2.4.
Middle: Histogram of continuous distribution, equal class width. *Right*:
Histogram for unequal class spacing.

We also may select *classes of unequal width*. For instance, we may pool
the largest values in a class comprising the porosities between 26 and
29%.

From now on, the following sign convention will be used: the small
letter f for absolute frequency, the letter q for relative frequency, and the
sign $q\%$ for 100 times q, that is for frequency by percentage.

	Equal class spacing				*Unequal class spacing*		
Class	Dash list	f	$q\%$		Class	f	$q\%$
19–20%	//	2	3.51		19–21%	3	5.26
20–21	/	1	1.75		21–22	9	15.79
21–22	╫╫ ////	9	15.79		22–23	17	29.82
22–23	╫╫ ╫╫ ╫╫ //	17	29.82		23–24	11	19.30
23–24	╫╫ ╫╫ /	11	19.30		24–25	7	12.28
24–25	╫╫ //	7	12.28		25–26	5	8.77
25–26	╫╫	5	8.77		26–29	5	8.77
26–27	/	1	1.75				
27–28	///	3	5.26		Sum	57	99.99
28–29	/	1	1.75				
Sum		57	99.98				

To represent these frequency tables graphically we draw a *histogram*.
This is constructed by plotting a series of contiguous rectangles whose
base length corresponds to the width of the class interval and whose area
corresponds to the frequency of the class (Figure 1). Thus, the total area
of all rectangles corresponds to the total number N of items if absolute
frequencies f are plotted, and to 100% (or unity respectively) if relative

frequencies are used. To indicate the scale, we draw separately a square whose area is 5 or 10% of the total histogram area.

In the situation where all classes are of equal width, the height of each histogram rectangle is proportional to f or q or $q\%$ respectively. This does not hold for unequal spacing! Thus, to avoid confusion, unequal spacing should be only used if absolutely necessary.

CHAPTER 3

Distribution Curves, Cumulative Frequency Curves, Special Representations, and Logarithmic Class Intervals

For large samples of a continuously varying property, rather small class intervals can be selected. In the limiting situation—infinitesimally small class widths and a sample of infinite size—the histogram of the relative frequencies becomes the *distribution curve* of the numerical variable x. Typical examples of smooth distribution curves are shown in Figure 2. The area between the distribution curve and an interval on the x-axis (with the endpoints x_1 and x_2) represents the relative frequency of the class ranging between x_1 and x_2 (Figure 2). If x_1 corresponds to the left endpoint of a continuous distribution curve and x_2 to the right endpoint the area is unity (100%). When x_1 and x_2 coincide, the area shrinks to zero: the relative frequency of a single value of a continuously distributed variable is always zero.

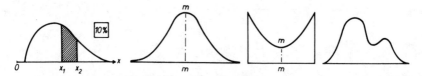

FIGURE 2. Some typical distribution curves; $m-m$: axis of symmetry. See text for further comments.

This condition does not hold for discrete distributions where x can assume only a finite number of discrete values (Example: a value of 2.4 is not possible if x is the number of garnet grains in a fine-sand fraction). For a discrete variable there exists no distribution curve but only a finite number of points which correspond to the different relative frequencies and the allocated x-values. Thus, there is a fundamental difference between discrete and continuous statistical variables. A distribution may be partly discrete and partly continuous. However, such situations are unusual in the geosciences.

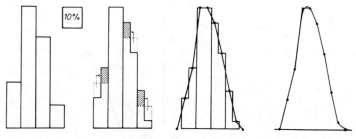

FIGURE 3. Histogram smoothing.

A distribution curve will be either asymmetrical or symmetrical with respect to some symmetry axis $m-m$ (Figure 2). The curve may have one or more maxima and they may be located anywhere. In the geosciences, many distribution curves have only one peak, so that the occurrence of two maxima in a sample distribution may indicate that two different populations with different maxima were blended in the past by some geological or biological process.

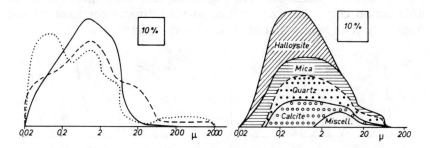

FIGURE 4. *Left-hand figure*: Grain-size distribution curves. *Full line*: red mud from Meteor station 290. *Dashed line*: red mud M.st. 251. *Dotted line*: blue mud M.st. 235 (Correns, 1939, p. 172). *Right hand figure*: Septarian clay from Malliss. Mineral composition and grain-size distribution (Correns, 1939, p. 178).

Strictly speaking, we never know exactly what the distribution curve of an empirical variable is because all samples are of finite size. Nevertheless, it may be advisable to smooth a step curve histogram as indicated in Figure 3. It goes without saying that this process of area shifting introduces an element of ambiguity. Smoothing, however, may be advantageous from the point of view of representation. A typical example is the superposition of several distribution curves as shown for instance in Figure 4.

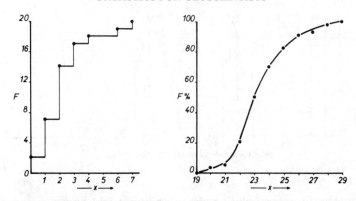

FIGURE 5. Cumulative frequency distribution curves (ogives). *Left-hand figure*: Distribution is discrete. *Example*: At seven locations, *x* is either equal to zero or to unity. *Right-hand figure*: Distribution is continuous. *Example*: Approximately 51% of all measured porosities are not larger (i.e., equal or smaller) than 23%.

By plotting the area under the distribution curve between *x* and its smallest value, the so-called *cumulative frequency distribution curve* (*ogive*) will be obtained. The ogive is less illustrative than the distribution curve, but can be used to read directly the total relative frequency of all values that are smaller (or larger) than *x*.

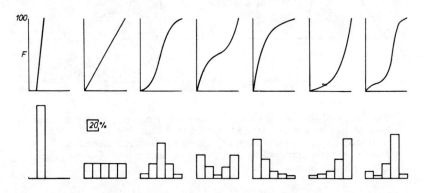

FIGURE 6. Typical distributions and corresponding cumulative curves.

Example 3.1   The tabulation of cumulative frequencies

The left-hand table shows the cumulative frequency *F* corresponding to Example 2.4, the right-hand one for Example 2.5. The corresponding ogives are represented in Figure 5. Note that the ogive of a discrete distribution is always a step curve because the cumulative frequency only changes for a finite number of values of x. Thus, never try to smooth the ogive of a discrete distribution as has been done previously.

x	f	F
0	2	2
1	5	7
2	7	14
3	3	17
4	1	18
5	0	18
6	1	19
7	1	20

x	$q\%$	$F\%$
19		0
20	3.51	3.51
21	1.75	5.26
22	15.79	21.05
23	29.82	50.87
24	19.30	70.17
25	12.28	82.45
26	8.77	91.22
27	1.75	92.97
28	5.26	98.23
29	1.75	99.98

In many instances distributions are represented as sectors of one or several circular disks (Figure 7 *left*). Such illustrations look useful and illustrative in maps. Corresponding *circular representations* are used to indicate the *distribution of vectors*, that is the distribution of physical entities characterized by both direction and numerical value, and visualized as arrows. The most important situations—the *vector histogram* and the projection of spatially distributed vectors on the horizontal plane—also are shown and explained briefly in Figure 7.

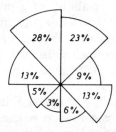
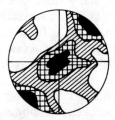

FIGURE 7. Circular distributions. *Left-hand figure*: Distribution of heavy minerals. Area of each sector is proportional to relative frequency of corresponding mineral. *Middle*: Vector histogram indicating frequencies of directions of vectors assembled in plane. *Right-hand figure*: Spatial frequency distribution of direction of vectors pointing upwards. Endpoint of each vector coincides with center of figure. Each arrowhead is projected on plane of map. Shading indicates relative frequency of direction. There are three maxima indicated by black areas.

Another special graphic representation of interest will be obtained when approximating the relative frequency distribution by a histogram with three class intervals of equal width. By using triangular coordinate paper (Figure 8), such distributions can be represented as a single point.

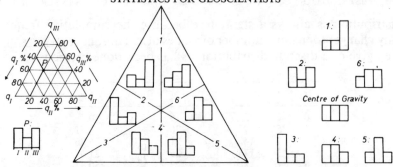

FIGURE 8. Representation of distributions on triangular coordinate paper. Triangular coordinates are shown at left. Point P corresponds to frequencies $f_I\%=40$, $f_{II}\%=20$, and $f_{III}\%=40$. (*Convention*: Draw straight line through point P parallel to right side of triangle until it meets triangle's left side. Read $f_I\%=40$. Then draw line through P parallel to left side of triangle until it meets triangle's base. Finally draw line through P parallel to triangle's base until it meets triangle's right side.) Other parts of Figure 8 indicate histograms corresponding to various points of triangle. There are bell-shaped distributions, U- and L-shaped histograms, etc.

To visualize the variability of evolutionary processes through time, an array of "snapshot" distribution curves may be plotted as indicated in Figure 33 of Chapter 13. Such a sequence of curves looks similar to a mountain chain where a ridge may branch (indicating the splitting off of a new evolution form) or break off (indicating extinction or migration of the considered biological population (Weigelt, 1950; Grabert, 1959; Bettenstaedt, 1962, p. 428–429).

In many instances the variability of a property is large, and small changes of the variable x for low values of x may be as important as large changes of x for large x-values. Typical examples are the grain-size distributions of various sediments. In such an instance the frequency distribution of log x rather than the distribution of x should be plotted by selecting *logarithmic class intervals.*

This nonlinear approach permits the investigation of extremely inhomogeneous populations, and relative changes of low values become as conspicuous as corresponding relative changes for large values. Obviously, the shape of a log-distribution is different from the shape of the corresponding initial distribution, especially with respect to symmetry properties and skewness (Figure 9). The log approach cannot be used if x can have a value of zero, for log x tends to minus infinity when x approaches zero.

Logarithmic representations are favored especially in the petrographical sciences. According to a proposal of W. C. Krumbein, the variable log x rather than the variable x should be considered as the relevant petrographic quantity. x has a definite physical meaning such as

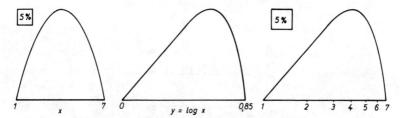

FIGURE 9. Log representations. *Left-hand figure*: Symmetrical distribution of variable *x*. *Middle*: Corresponding asymmetrical distribution of $y=\log x$ (log $1=0$, log $7=0.85$). *Right-hand figure*: Same distribution as middle figure. Abscissa, however, is labeled with *x*-values.

designating grain-size diameters, whereas log *x* is something more abstract. Yet Krumbein's approach simplifies the statistical treatment and evaluation considerably and has been accepted widely.

Two types of equally spaced logarithmic class intervals are mainly in use. The first is based on logarithms to the base of 10 (in petrography also termed "zeta scale"), the second on logarithms to the base of 2, a logarithmic derivative of the "Wentworth scale"* (in petrography termed "phi scale") if the variable *x* represents grain-size diameters *d* measured in millimeters (mm).

Zeta scale

Zeta$=\log_{10}x$

zeta:	-2	-1	0	1	2	3
x:	0.01	0.1	1	10	100	1000

Phi scale

Phi$=-\log_2 d$ $\qquad d=2^{-Phi}$

Φ:	-6	-5	-4	-3	-2	-1	0	1	2	3	4	5	6	7	8	9	10
d:	64	32	16	8	4	2	1	$\frac{1}{2}$	$\frac{1}{4}$	$\frac{1}{8}$	$\frac{1}{16}$	$\frac{1}{32}$	$\frac{1}{64}$	$\frac{1}{128}$	$\frac{1}{256}$	$\frac{1}{512}$	$\frac{1}{1024}$ mm

Most sedimentologists use "Φ-grades".

*Wentworth is a geometric scale. $\log_2$ (Wentworth classes) is a logarithmic scale, Φ W.C. Krumbein.

CHAPTER 4

Averaging

When studying the variability of a numerical property, one usually is interested first in some type of average that characterizes the property in the mean.

Notations

x_1, x_2, x_3, ... x_n: sampled values of the variable x. For a continuously distributed variable, x_1, x_2, etc., may be the midpoint values of the selected classes.

f_1, f_2, f_3, ..., f_n: absolute frequencies of sampled values. If x_i represents the midpoint value of class i, then f_i is the number of sampled values contained in class i.

N: sample size $f_1 + f_2 + f_3 + ... + f_n$.

q_1, q_2, q_3, ... q_n: relative frequencies where $q_i = f_i/N$.

Averages

The mode M: The value of x with the largest frequency; the value of x corresponding to the highest peak of the distribution curve (histogram).

The median Q_2: For $x = Q_2$, the cumulative frequency is equal to 50%.

The mean $\bar{x}$ (read "x bar"):

$$\bar{x} = (f_1 x_1 + f_2 x_2 + f_3 x_3 + ... + f_n x_n)/N$$

$$\bar{x} = q_1 x_1 + q_2 x_2 + q_3 x_3 + ... + q_n x_n$$

The geometric mean G: The Nth root of the product of the numbers z_1, z_2, z_3, ..., z_n where z_i is the f_ith power of x_i. The geometric mean is not defined if at least one x-value is negative. *Example*: $x_1 = 2$, $x_2 = 8$, $f_1 = f_2 = 1$. G = square root of $16 = 4$. The logarithm of G is equal to the mean of the corresponding log x-distribution:

$$\log G = (f_1 \log x_1 + f_2 \log x_2 + f_3 \log x_3 + ... + f_n \log x_n)/N$$

The harmonic mean H:

$$H = N/(y_1 + y_2 + y_3 + \ldots + y_n) \qquad \text{where } y_i = f_i/x_i.$$

Example: $x_1 = 2$, $x_2 = 8$, $f_1 = f_2 = 1$. $H = 2/(0.5 + 0.125) = 3.2$

The *vector mean* is discussed at the end of the chapter. There are other types of averages as for instance the geometric-harmonic mean. They are, however, seldom used in statistics.

	Summary of the most important averages
$x_1, x_2, x_3, \ldots, x_n$	sampled values of the variable x (continuous example: midpoint values of selected classes)
$f_1, f_2, f_3, \ldots, f_n$	absolute frequencies
$q_1, q_2, q_3, \ldots, q_n$	relative frequencies: $q_i = f_i/N$, $q_i\% = 100 f_i/N$
$f_1 + f_2 + f_3 + \ldots f_n = N$	sum of absolute frequencies = sample size N
$q_1 + q_2 + q_3 + \ldots + q_n = 1$	sum of relative frequencies = 1
$q_1\% + q_2\% + \ldots + q_n\% = 100$	sum of relative frequencies (%) = 100 Often, absolute frequencies are not known (example: grain-size distribution of a sand)
Mode M	the most abundant value
Median Q_2	cumulative frequency is equal to 50% at $x = Q_2$ (Q_2 may not exist for a discrete distribution)
Mean $\bar{x}$	(read "x bar") $\bar{x} = \{f_1 x_1 + f_2 x_2 + \ldots + f_n x_n\}/N = \{q_1 x_1 + q_2 x_2 + \ldots + q_n x_n\}$
Geometric mean	$G = {}^N\!\sqrt{x_1^{f_1} \cdot x_2^{f_2} \cdot x_3^{f_3} \ldots x_n^{fn}}$ (all $x \geq 0$)
Harmonic mean	$H = N/\left\{\dfrac{f_1}{x_1} + \dfrac{f_2}{x_2} + \ldots + \dfrac{f_n}{x_n}\right\}.$

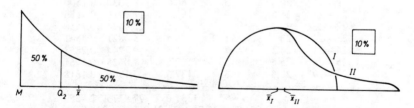

FIGURE 10. Some distributions and their averages. *Right-hand figure*: Modes of two distributions coincide.

The harmonic mean is never larger than the geometric mean, and the latter is never larger than $\bar{x}$. For every symmetrical distribution curve of otherwise arbitrary shape, the median and the mean coincide. For asymmetrical distributions, $\bar{x}$ and Q_2 usually differ. For nearly bell-shaped curves with slight to medium skewness, the median is approximately equal to one-third of $\bar{x}+\bar{x}+M$.

	Relations between averages
$H<G<\bar{x}$	$(H=G=\bar{x}$ if and only if $x_1=x_2=x_3=\ldots=x_n)$
$\bar{x}=Q_2$	if the distribution curve is symmetrical (usually $\bar{x}\neq Q_2$ for nonsymmetrical curves)
$3Q_2=2\bar{x}+M$	approximate formula for bell-shaped distribution curves of moderate skewness
	$\log G=\{f_1\log x_1+f_2\log x_2+\ldots+f_n\log x_n\}/N$ The mean of the log $x-$distribution is equal to the logarithm of the geometric mean. (Practical formula for calculating $G)$

Example 4.1   The calculation of averages

Distribution: Example 2.5
$M\ \ =22.5$ (Figure 1)
$Q_2\ \ =23.0$ (Figure 5)

Class	x	f	$x \cdot f$	$f \cdot \log x$	f/x
19—21	20.0	3	60.0	3.9030	0.1500
21—22	21.5	9	193.5	11.9916	0.4186
22—23	22.5	17	382.5	22.9874	0.7556
23—24	23.5	11	258.5	15.0821	0.4681
24—25	24.5	7	171.5	9.7244	0.2857
25—26	25.5	5	127.5	7.0325	0.1961
26—29	27.5	5	137.5	7.1965	0.1818
		$N=57$	1331.0	77.9175	2.4559

$\bar{x}=1331/57=23.35$ $G=23.28$ $H=23.21$

What type of average is the most appropriate one? This question will be answered by and large in the following survey.

The mode is easy to determine, easy to take in at a glance but somewhat unsatisfactory for bimodal and multimodal distributions, that is for distributions with several peaks. *Main disadvantage*: M is an isolated rather than a global property of a distribution; M refers only to one specific x-value. On the other hand, it might be important to know where the frequency accumulates.

The median is easy to determine. It is rather insensitive with respect to the tails of the distribution. This can be an advantage or a disadvantage. Q_2 has the smallest mean deviation of all possible averages (the term "mean deviation" will be defined in Chapter 5).

Main disadvantage: In general, the Q_2-value of a sample S deviates more from the Q_2-value of its universe U than the $\bar{x}$-value of S differs from the $\bar{x}$-value of U. The median is favored by sedimentologists who do not use a computer.

The mean in general is more representative of the sample than Q_2. The professional statistician prefers to use the mean because most statistical tests for averages refer to it.

The geometric mean: Its use is restricted mainly to three aplications.

(1) Interpolation of geometric progressions or exponential curves. *Example*: The geometric mean of the numbers 10 and 1000 is 100; the sequence 10, 100, 1000 represents a geometric progression.

(2) Let a, b, c be the lengths of the main axes of a grain. Then the third root of the product abc is an appropriate average grain diameter because the volume of an ellipsoid is proportional to abc.

(3) The geometric mean is used occasionally in petroleum reservoir engineering.

The harmonic mean: seldom used except in petroleum reservoir engineering. Note that both G and H vanish if at least one x-value is equal to zero. Both G and H tend to favor small x-values and to disfavor large values!

Summary of properties of the averages	
M	isolated property of limited value but easily determined; rather inappropriate for bimodal and multimodal distributions.
Q_2	easily determined and rather insensitive to the shape of the tails of the distribution curve. It depends on the situation whether the last property is an advantage or not. For large samples from a normal population, the mean has a smaller variance than the median (variance: see next chapter; see also Chapter 10).
$\bar{x}$	the mean is the most important average. Most distributions are characterized sufficiently by the mean and the variance.
G	the radius of a grain with the main axes $a, b,$ and c is characterized best by the third root of the product abc.
H	seldom used in statistics.

Example 4.2   The theoretical center of glacial deposits, TCD

The glacial deposits in northwestern Germany generally originated from Scandinavia. They usually contain pieces of rock material of Scandinavian origin which can be located rather closely to source. The center of all initial locations can be estimated by calculating the mean geographical longitude λ and the mean geographical latitude $\bar{\varphi}$ of all positions from which transportation started. The calculated center corresponds closely to the central point of the region from which the rock particles were transported by glaciers.

Fictitious sample from a ground moraine

f: number of rock pieces with diameters between 6 and 63 mm,
λ, φ: geographical coordinates of the initial locations of the debris.

Rock	λ	φ	f	$f \cdot \lambda$	$f \cdot \varphi$
Alminding granite	15.0	55.1	26	390	1433
Brevik diabase	14.7	57.6	12	176	691
Fagerhult tuff	15.6	57.2	33	515	1888
Labrador porphyrite	24.3	60.8	51	1239	3101
Tessini-sandstone	16.3	56.4	8	130	451
Sums			130	2450	7564

2450/130=18.8   7564/130=58.2   TCD:λ = *18.8°* $\bar{\varphi}$ = *58.2°*

The theoretical center of glacial deposits was introduced by Luettig (1958). According to his investigations, the TCDs of ground moraines in northwestern Germany are in a well-defined region of the moraines are located in an area of about 5000 square miles. This permits in many situations a determination of the age of the sample.

Circular distributions. These distributions constitute an important class of distributions in the geological sciences. For example, samples describing the variability of angular variables such as strike and dip, orientation of mineral grains, and angular parameters related to bedding planes, faults, fissure systems, cross bedding, ripple structures, etc., are circular frequency distributions.

From the geometrical point of view, there are essentially four categories: The differing direction (angle) either may be confined to a fixed plane or distributed in three-dimensional space; a direction and its opposite direction may be equivalent or not.

Obviously, the simplest representation of an angular variable that is most appealing to the eye is shown at the left-hand of Figure 12. The corresponding histogram representations (also Figure 12) are not unique: The shape of the histogram and the mean value of the

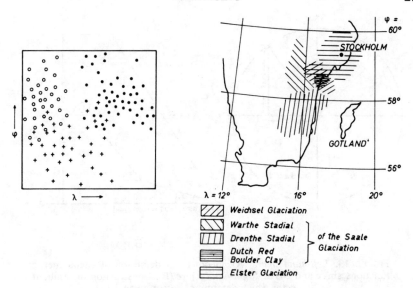

FIGURE 11. *Left-hand figure*: Typical distribution of TCDs for three different ages. Corresponding regions overlap. *Right-hand figure*: TCD-regions for ground moraines located in areal Hanover–Westphalia–Ruhr valley–Netherlands (Luettig, 1958).

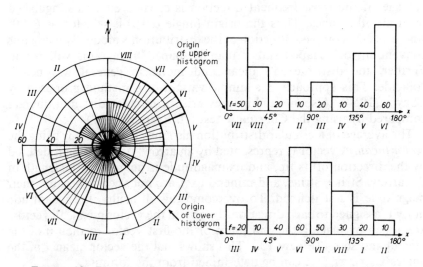

FIGURE 12. Representation of circular distributions. *Left-hand figure*: Segmented circular disk. All lines designating direction originate from center of disk. Frequency of each class (represented by segment) is proportional to lengths of corresponding lines. Distribution is symmetric because each direction is equivalent to its opposite direction. (*Example*: Orientation of *c*-axes of calcite in rock.) *Right-hand Figure*: Corresponding histograms.

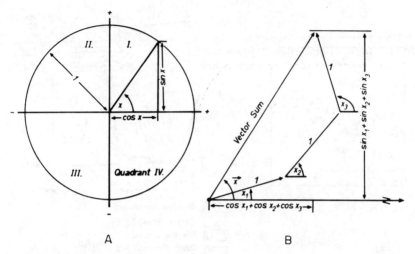

A   B

FIGURE 13. *Left-hand figure* (A): Geometrical definition of trigonometric functions sin x and cos x. *Right-hand figure* (B): Construction of resultant vector which determines vector mean.

corresponding frequency distribution depend in general on the selection of the histograms origin. In the upper histogram of Figure 12, the counting of angles begins at northeast, for the lower histogram at southeast. The two representations are completely different. The ambiguity vanishes if the most frequent direction is considered as distinguished from all other ones. Thus the origin (angle $\alpha=0$) is selected as $\theta-90°$ where θ is the compass direction of the distribution's mode. By using this convention, bell-shaped rather than U-shaped histograms will result. Further, the distribution's mean and mode in general will nearly coincide. This approach is a simple variant of a method proposed by Chayes (1954). Also no ambiguities arise if circular distributions are compared by using the Chi-square test (cf. Example 11.7).

The average of a circular distribution also may be defined by using the *vector mean*. A vector is represented by an arrow; its direction is defined by the direction of its tip, and its magnitude is defined by the length of the arrow. Strike, striae, and mineral axes are not vectors because their magnitude is not defined. To overcome this difficulty, the magnitude unity is assigned to each direction. To obtain the vector mean, all vectors are added by placing the end point of the first vector to the tip of the second and so on (Figure 13.). This shows that the vector mean $\bar{x}$ of the angles $x_1, x_2, x_3, \ldots,$ can be determined from the formula

$$\tan \bar{x} = (\sin x_1 + \sin x_2 + \ldots + \sin x_n)/(\cos x_1 + \cos x_2 + \ldots + \cos x_n)$$

It may happen that the resultant vector vanishes, that is that the end point of the first vector and the tip of the last one coincide. In this situation the vectors form the sides of a closed polygon. Then the vector mean does not exist and the calculation results in $\tan\bar{x}=0/0$ which is indeterminate. It also may happen that the end-point and the tip of the resultant vector are close to each other. In this situation a slight change of the direction of some vector may have a large effect on $\bar{x}$. Hence its use is not recommended too strongly. For additional literature about the vector mean see Krumbein (1939), Scheidegger (1965), Curray (1956), Pincus (1956), Raup and Miesch (1957), and Steinmetz (1962).

CHAPTER 5

Variance, Standard Deviation, Skewness, and Kurtosis. Moments

There are essentially four approaches by which the variability of quantitative data may be characterized. In the following definitions, each distribution is characterized by n numbers $x_1, x_2, x_3, \ldots, x_n$ and the corresponding absolute frequencies $f_1, f_2, f_3, \ldots f_n$. The number N is the sum $f_1 + f_2 + f_3 + \ldots f_n$.

The range. The variability indicated by the difference between the smallest x-value, x-min and the largest x-value, x-max:

$$L = x\text{-max} - x\text{-min}$$

is termed the *range of the distribution.* This simple approach to variability is used widely in the older paleontological literature. The number L is a measure of the "distance" between the "tails" of a distribution. The range of a sample, however, in general may not be representative of the range of the total population from whch the sample has been drawn and as a consequence, the range is seldom used now.

The mean deviation (about the mean $\bar{x}$) is defined by the expression

$$(f_1|x_1 - \bar{x}| + f_2|x_2 - \bar{x}| + \ldots + f_n|x_n - \bar{x}|)/N$$

The symbol $|x|$ designates the *absolute value* of x. *Examples:* $|+3| = 3$, $|-1| = 1$, $|0| = 0$. The use of the mean deviation is restricted to the older literature.

If in the given expression the mean, $\bar{x}$, is replaced by the median, Q_2, the *mean deviation about the median* will result. Q_2 and the mean deviation about Q_2 are a fair representation of a sample and its population.

The variance and its square root, the *standard deviation* or *root mean square deviation*, are the measures of variability that are used by the statistician. The variance of a finite population, σ^2, is defined by the formula

$$\sigma^2=[f_1(x_1-\bar{x})^2+f_2(x_2-\bar{x})^2+ \ldots +f_n(x_n-\bar{x})^2]/N$$

In most applications, it is sufficient to characterize a distribution by its mean, $\bar{x}$, and its variance, σ^2. The variance of a sample is designated by the symbol s^2. One always should distinguish carefully between the *population variance*, σ^2, and the *sample variance*, s^2. If s^2 is known, the best estimate (*unbiased estimate*) of the population variance, designated $\hat{s}^2$ is

$$\hat{s}^2=Ns^2/(N-1).$$

The understanding of variance and mean deviation are easy to grasp. The difference $x_1-\bar{x}$ is the deviation of x_1 from the mean $\bar{x}$. Thus, variance and mean deviation are the means of deviations or squares of deviations respectively. Both variance and mean deviation take into account all values of the distribution whereas the range reflects only the distance between the tails of a distribution.

Summary	
Range	$L=x\text{-max}-x\text{-min}$
Mean deviation ($a=\bar{x}\ or\ a=Q_2$)	$\{f_1|x_1-a|+f_2|x_2-a|+ \ldots +f_n|x_n-a|\}/N$
Variance: Population variance ($N=$size of population)	$\sigma^2=\{f_1(x_1-\bar{x})^2+f_2(x_2-\bar{x})^2+ \ldots +f_n(x_n-\bar{x})^2\}/N$
Sample variance ($N=$size of sample)	$S^2=(f_1(x_1-\bar{x})^2+f_2(x_2-\bar{x})^2+ \ldots +f_n(x_n-\bar{x})^2)/N$
Unbiased estimate of σ^2 ($N=$size of sample)	$\hat{s}^2=s^2\cdot N/(N-1)$
Standard deviation, root mean square deviation	positive square root of variance

The defining equation of the variance may be recast into the useful equivalent form

$$\sigma^2=\frac{1}{N}\{f_1x_1^2+f_2x_2^2+ \ldots +f_nx_n^2\}-\bar{x}^2$$

Example 5.1   The variability of the distribution from
Example 2.4

x	f	xf	$\|x-\bar{x}\|$	$f\|x-\bar{x}\|$	$(x-\bar{x})^2$	$f(x-\bar{x})^2$	fx^2
0	2	0	2.25	4.5	5.06	10.12	0
1	5	5	1.25	6.25	1.56	7.8	5
2	7	14	0.25	1.75	0.06	0.42	28
3	3	9	0.75	2.25	0.56	1.68	27
4	1	4	1.75	1.75	3.06	3.06	16
5	0	0	2.75	0	7.56	0	0
6	1	6	3.75	3.75	14.06	14.06	36
7	1	7	4.75	4.75	22.56	22.56	49

$$
\begin{array}{llll}
\quad 20 \quad 45 & \qquad\qquad 25.00 & \qquad\qquad 59.7 \quad 161 \\
\bar{x}=45/20=2.25 & 25/20=1.25 & s^2=59.7/20= \\
& & (161/20)-2.25=2.99 \\
& & s=+\sqrt{2.99}=1.73
\end{array}
$$

$$x_{min}=0, \quad x_{max}=7, \quad L=7.$$

Example 5.2   The variance of flow directions in a fossil
river bed

Suppose that the total area of a fossil river bed is divided into areas of
equal size. For each area, the directions of strike of beds with cross-
stratification are recorded and the vector means and the variances are
calculated. A map representing all vector means as arrows reveals the
main flow direction of the fossil river. A map of the lines of equal
variance shows the stability or instability of the main flow directions. A
frequent change of direction may indicate a shifting of the mouth of a
fossil tributary. In the simplest situation, there are two vector maxima
that characterize the directions of flow of the main river and the
tributary respectively (Figure 14).

Example 5.3   The variance of biological forms

The variance within a biological population may depend on different
factors and may indicate the phylogenetic situation of the considered
form. It may be of interest to investigate the dependence of the variance
on location. The red deer (*Cervus elaphus*) occurs in several sizes that
seem to depend on the climate (continental or maritime respectively).
Does the variance of size also depend on climate and location?

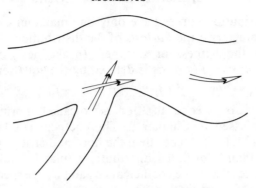

FIGURE 14. Flow directions in fossil riverbed. Intersecting arrows may indicate mouth of tributary.

Example 5.4   The range of variability of the beetle Carabus *in relation to climate*

Rensch (1947, p. 44) compares the variabilities of the lengths, breadths, etc., of thirty-one male beetles (*Carabus coriaceus coriaceus*) from Central Europe with the variabilities of thirty-one beetles (*Carabus coriaceus cerisyi*) from Greece and Turkey. He considers the differences as significant and draws far-reaching genetic conclusions. The results however, do not seem to be justified because the differences in variability usually are fractions of millimeters and not representative because the sample size ($N = 31$) is too small. In such a situation the variances and test of significance of the difference must be calculated by Snedecor's F-test as will be explained in Chapter 11.

The mean and the variance of a distribution do not indicate whether the distribution curve is symmetrical. To determine the *skewness* of a distribution, the dimensionless *moment coefficient of skewness* may be calculated.

$$\alpha_3 = (f_1(x_1 - \bar{x})^3 + f_2(x_2 - \bar{x})^3 + \ldots + f_n(x_n - \bar{x})^3)/Ns^3.$$

The coefficient α_3 is zero if, and only if, the distribution is symmetrical. To some extent, skewness also is indicated by *Pearson's first or second measure of skewness*

$$P_1 = (\bar{x} - M)/s$$

$$P_2 = 3(\bar{x} - Q_2)/s.$$

P_1 vanishes if the mean $\bar{x}$ and the mode M coincide. P_2 vanishes if the mean and the median Q_2 are equal. In general, P_1 is useless if the distribution has at least two maxima. P_2 is always zero if the distribution is symmetrical.

Many distribution curves have only one maximum. Then it may be useful to characterize the *kurtosis* of the distribution. The kurtosis is a measure for the flatness or steepness (peakedness) of a curve. *The moment coefficient of kurtosis* is defined by the equation

$$\alpha_4 = \{f_1(x_1 - \bar{x})^4 + f_2(x_2 - \bar{x})^4 + \ldots + f_n(x_n - \bar{x})^4\}/Ns^4.$$

The coefficient α_4 is equal to 3 for the so-called "normal distribution" which is discussed in Chapter 9. The coefficient is larger than 3 for distributions that are steeper than the normal distribution (leptokurtic); it is smaller than 3 for flatter distributions (platykurtic).

The variance σ^2 and the coefficients α_3 and α_4 are special situations of the *moments of a distribution*. The *moments about the mean* are defined by the equations

$$\mu_r = \{f_1(x_1 - \bar{x})^r + f_2(x_2 - \bar{x})^r + \ldots + f_n(x_n - \bar{x})^r\}/N$$
$$\alpha_r = \mu_r/\sigma^r$$
$$r = 2, 3, 4, \ldots \qquad \text{(population)}$$
$$m_r = \{f_1 - \bar{x})^r + f_2(x_2 - \bar{x})^r + \ldots + f_n(x_n - \bar{x})^r\}/N$$
$$a_r = m_r/s^r$$
$$r = 2, 3, 4, \ldots \qquad \text{(sample)}$$
$$\mu_2 = \sigma^2$$
$$m_2 = s^2$$
$$\alpha_2 = 1$$
$$a_2 = 1$$

Most distributions are determined uniquely by the mean and their moments. In most practical situations it is sufficient to characterize a distribution by its mean and its variance and a few moments much as α_3 and α_4.

When a distribution is split into several classes, the actual sample values may be replaced by the midpoint values of the classes into which the samples have been grouped.

Example:

Sample values: 0.1, 0.1, 0.3, 0.5, 0.6, 0.8, 1.2, 1.3, 1.5, 1.5, 1.9
Class 1: Range: $0-1$; $x=0.5$; $f=6$
Class 2: Range: $1-2$; $x=1.5$; $f=5$

This simplification introduces errors when calculating the moments. These errors can be eliminated partly by applying *Sheppard's corrections*:

$\bar{x}$: no correction
s^2: subtract $c^2/12$ (c=mean length of the class intervals)
m_3: no correction
m_4: subtract $c^2s^2/2$ and add $7c^4/240$

These corrections tend to overcompensate and can be rather inaccurate if the class interval lengths are unequal and if the tails of the distribution are short. In general, Sheppard's corrections can be recommended no longer. All moments can be calculated easily and exactly using a computer program. Thus, there is no need to concentrate the sample values on the classes' midpoints.

CHAPTER 6

The Characterization of Grain-size Distributions

Most sedimentologists describe grain-size distributions by selected points of the cumulative frequency distribution, F. Every pair "grain size x, corresponding cumulative percentage weight frequency F" is termed a *quantile*. The following conventions are mainly in use:

	Quantiles Q_i		Deciles D_i		Percentiles P_i	
x	F	x	F	x	F	
Q_1	25%	D_1	10%	P_1	1%	
Q_2	50%	D_2	20%	P_2	2%	
Q_3	75%	....		....		
		D_9	90%	P_{99}	99%	

Q_1 is the first quantile; Q_2 is the second quantile; P_1 is the first percentile, and so on. Quantiles are used widely to characterize distributions by parameters that correspond roughly to the moments of mathematical statistics:

Average:	Median	Q_2
Variability:	Quantile deviation	$\frac{1}{2}(Q_3-Q_1)$
	Trask's sorting coefficient	$(Q_3/Q_1)^{\frac{1}{2}}$
Skewness:	Quantile coefficient of skewness	$(Q_3-2Q_2+Q_1)/(Q_3-Q_1)$
	Trask's coefficient of skewness	Q_1Q_3/Q_2^2

Trask's coefficient of skewness is equal to unity if $\log Q_2$ is the arithmetic mean of $\log Q_1$ and $\log Q_3$:

$$Q_1Q_3/Q_2^2 = 1 \text{ if and only if } \log Q_2 = \tfrac{1}{2}(\log Q_1 + \log Q_3).$$

This condition is satisfied if $\log x$, the logarithm of grain size, is distributed symmetrically. Inman (1952), McCammon (1962), and Folk (1966) define distribution parameters by using deciles and percentiles. Sharp and Fan (1963) introduced a sorting index that takes into account all grain-size classes.

32

Coefficients related to quantiles have the following properties in common:

1. They are calculated easily.
2. They take into account only a few points of the distribution curve. For instance, there exists an infinity of skewed distributions such that Trask's coefficient of skewness is equal to unity.
3. They are less representative than the mean and the moments. For example, the median Q_2 of a sample is not as representative of the population as the sample mean. The deciles and percentiles near the tails of a distribution such as D_1, D_9, P_1, P_3, P_{99}, etc., are not representative. Chances are high that, say, the value of P_{97} of a sample is different from the corresponding population value.
4. There are no known methods to estimate the quantiles of a population from the quantiles of a small sample.

There is, nevertheless, a strong argument for the use of the quantiles Q_1, Q_2, and Q_3. Their values are independent of the distribution of the finest fractions which in many instances remain unknown (Folk, 1966).

CHAPTER 7

Statistical Analysis of Inhomogeneity

A system is termed *homogeneous* if it is uniform, that is if it has the same properties at every point or volume element. Most systems such as geological bodies, rocks and minerals may be homogeneous with respect to one property but *inhomogeneous* with respect to another property. For instance, a pure quartz sand composed of grains of different size is homogeneous with respect to mineral composition but inhomogeneous with respect to grain size. Most geological bodies and most rocks are inhomogeneous in many respects, and the spatial distribution of their inhomogeneities determines to a large extent or completely their texture.

In the crystalline state the atoms are located at different points in a lattice arrangement: the inhomogeneities are *ordered*. This situation is dealt with in crystallography. In amorphous substances such as wax, tar, glasses, and many alloys, the atoms are either *randomly distributed* or *partly ordered*. Other examples of partial ordering or random spatial distribution are furnished by sequences of layers of different rocks or the arrangement of minerals in crystalline rocks.

Hierarchy of Inhomogeneity

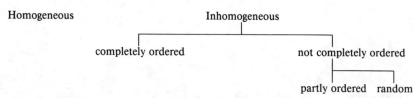

The investigation of partial ordering and random spatial distribution is simple if the analysis is confined to single directions (such as to the normal of a sequence of layers). Suppose there is an arbitrary *sequence* of elements, each element representing either property *a* or *b*. Each sequence of elements of same type is termed a *run*. For instance, the arrangement *aaabbaba* consists of eight elements and the five runs *aaa*, *bb*, *a*, *b*, *a*. The run *aaa* has the "*length*" 3 (=number of elements of a run), the run *bb* has length 2 and the other runs have length 1.

Correspondingly, the sequence *aacbabc* consists of three types of elements (*a*, *b*, *c*), the total number of elements is equal to seven, and the total number of runs is equal to six (*aa*, *c*, *b*, *a*, *b*, *c*).

The arrangement of the elements of a run may be ordered, randomly distributed, or partly ordered. In the latter situation the mean lengths of the run may be shorter or longer than the mean lengths of the corresponding random arrangement. (The mean length of the *a*-runs *aaa*, *a*, *a* is equal to $(3+1+1)/3 = 5/3$. The mean length of the runs *bb*, *b* is equal to $(2+1)/2 = 3/2$. The mean length of all runs is equal to $(3+1+1+2+1)/5 = 8/5$.)

A sequence of elements is *ordered* if a pattern repeats itself (*cyclic pattern*). *Example*: the pattern is *aabc*, the sequence is *aabcaabcaabcaab-caabcaabc*. To study *partial ordering* of a sequence of elements of two or more mutually exclusive types we may compare the *actual sequence* with a *reference cyclic pattern*. The comparison results in a *deviation pattern* which constitutes a measure of the partial ordering with respect to the reference sequence.

Example 7.1 *Investigating and quantifying a partial ordering*

Suppose we consider the arrangement of the minerals *a*, *b*, and *c* in a rock along some specified direction. The actual sequence is compared to the cyclic pattern ... *aabc* ... :

Actual sequence:	abbc	aabc	babc	aaba	aaac
Cyclic reference pattern:	aabc	aabc	aabc	aabc	aabc
Deviation sequence:	0100	0000	1000	0001	0010

0: actual element and reference element are of same type
1: corresponding elements are of different type

Obviously, the actual arrangement is not random but may be considered as close to the cyclic pattern *aabc* with deviation quota of 20% (four deviations out of twenty comparisons). There might exist another cyclic pattern with smaller deviation quota. The deviation sequence $010000...$ consists of two types of elements (0 and 1) and may be described and analyzed for randomness as shown next.

The deviation sequence of the last example or any other sequence of runs can be considered as a set of ordinary frequency distributions. For instance, the deviation sequence of Example 7.1 contains the following frequency distributions.

Element: 0		Element: 1		Elements: 0,1	
Run length	Frequency	Length	Frequency	Length	Frequency
1	2	1	4	1	6
2	1			2	1
6	2			6	2

The frequencies of all other lengths are zero. Each frequency distribution has a mean and a variance which may be compared with the corresponding parameters of a random arrangement of elements.

For random sequences which consist of two types of elements, say a and b, the following formulae hold:

$$n = \text{number of } a\text{s}$$
$$m = \text{number of } b\text{s}$$
$$k = n + m = \text{total number of elements}$$

Type	Number of runs	Mean length of all runs
a	$n(m+1)/k$	$k/(m+1)$
b	$m(n+1)/k$	$k/(n+1)$
a,b	$1+(2nm/k)$	$k/[1+(2nm/k)]$

Type	Variance of the number of runs
a	$mn(n-1)(m+1)/[kk(k-1)]$
b	$mn(m-1)(n+1)/[kk(k-1)]$
a,b	$2nm(2nm-k)/[kk(k-1)]$

Numerical Examples

Actual sequence: a bb a b aa $bbbb$ a b a bbb aa b a b a bb

$$n = 10 \ (10 \ a\text{s}) \qquad m = 15 \ (15 \ b\text{s}) \qquad k = 25$$

Frequency distributions:

	Element: a		Element: b		Elements: a,b
Length	Frequency	Length	Frequency	Length	Frequency
1	6	1	4	1	10
2	2	2	2	2	4
		3	1	3	1
		4	1	4	1

Distribution	Number of runs	Mean length
a-actual	8	1.25
a-random	6.4	1.5625
b-actual	8	1.875
b-random	6.6	2.273
a,b-actual	16	1.5625
a,b-random	13	1.9231

For a single sequence, we can determine the variances of run lengths but not the variance of the number of runs because there is just one number of runs for each type of elements (such as "8 a-runs"). Let us therefore consider another example with several sequences:

$$n=2 \qquad m=3 \qquad k=5$$

Sequence	Total number of runs
abbab	4
aabbb	2
babab	5
bbaab	3
ababb	4

Frequency distribution of the number of runs

Number of runs	Frequency
1	0
2	1
3	1
4	2
5	1

Mean number of runs		Variance of the number of runs	
Actual	Random	Actual	Random
3.8	3.4	1.08	0.84

To test the statistical hypothesis that two sequences do not differ significantly, we have to apply the methods of Chapter 11 to the distribution of runs of the two sequences. To do so we must compute and compare means and variances of run lengths. (For details see Chapter 11.) This method is straightforward, simple, and general; it can be applied to sequences which consist of any number of different types of elements (say twelve different rocks or nine different minerals or two types of minerals such as "dark" and "light colored").

To test a single sequence of randomness, the sequence must be compared to a corresponding random sequence.

Example 7.2   Testing a sequence of randomness

The sequence consists of four types of elements termed *a*, *b*, *c*, d.

Number of *a*s :20
Number of *b*s :12
Number of *c*s : 8
Number of *d*s :35
Total number :75

Put 75 tickets labeled *a* (20 times), *b* (12 times), etc., in a box and draw the tickets at random. There results a random sequence composed of 20 *a*s, 12 *b*s, etc. Calculate the means and variances of the random run lengths and apply the methods of Chapter 11. Obviously, it is more elegant to write a computer program which produces the integers 1 through 75 at random, each once. Then the integers 1 through 20 are interpreted as elements of type *a*, the integers 21 through 32 are interpreted as *b*s, etc.

Literature on the theory of runs: Miller and Kahn (1965, chapter 14), Wilks (1946, chapter 10). Wilks derives the theory of runs, Miller and Kahn are more interested in applications.

Example 7.3   Rock texture

The fabric of a rock may be described quantitatively by applying the theory of runs to the arrangement of the rock's mineral grains. For each type of mineral or each specified group of minerals the mean run length *L* is determined along parallel lines the distance of which is proportional roughly to the mean grain size. For a monomineralic rock, *L* is infinite (Figure 15A). A fabric is isotropic if *L* is independent of direction (Figure 15B). A fabric is anisotropic if *L* differs significantly for different directions (Figure 15C, *L*=1 and *L*=infinite respectively along the diagonals of the figure).

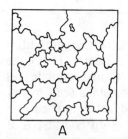
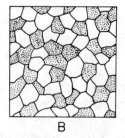
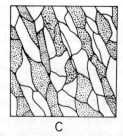

A   B   C

FIGURE 15. Mean length *L* of run of mineral grains as quantitative property of rock fabric. A: Monomineralic rock, *L*=infinite. B: Isotropic fabric, *L* does not depend on direction. C: Anisotropic fabric, that is *L* depends on direction. In the figure *L*=1 and *L*=infinite respectively along the diagonals.

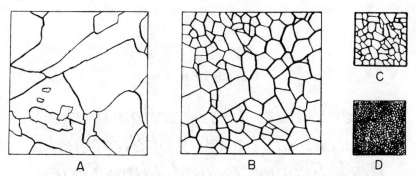

FIGURE 16. A: 10 grains per cm², B: 100 grains per cm ² (linear magnification: four times). C: 1000 grains per cm², D: 10000 grains per cm² (linear magnification: six times) Teuscher, 1933).

Runs should be counted along parallel lines with spacing D where D is the square root of the specific grain number, that is the number of grains per unit of area. The smallest grain-size fraction and large grains should be omitted when estimating the specific grain number.

CHAPTER 8

Elementary Introduction to the Statistical Analysis of Stratified Rock Sequences and Time Series

The topics of this chapter are concerned mostly with rather complex mathematical techniques that are beyond the scope of the text. Therefore, the discussion will be restricted to a few simple illustrations.

Example 8.1 Sequential analysis of a sedimentation cycle

We illustrate the technique by applying it to seven profiles with three cycles each. First, we will construct an "ideal profile" by comparing the seven different profiles. The letters a, b, c, d, e, f stand for different types of beds; the capital letter H indicates a bed that is missing with respect to an ideal sequence of layers constructed from the actual beds. The number of "hiata," that is the total number of occurrences of the letter H, should be kept as small as possible.

Profile:	1	2	3	4	5	6	7	Ideal profile	Number of deviations
	a	a	a	H	a	a	a	a	1
	b	b	b	b	b	b	b	b	0
	c	c	c	c	c	c	c	c	0
	H	a	H	a	a	H	a	a	3
	d	d	d	d	d	d	d	d	0
	e	e	e	e	e	e	e	e	0
	f	f	f	f	f	f	f	f	0
									(end of first cycle)
	a	a	a	a	c	a	a	a	1
	H	H	b	b	H	b	H	b	4
	c	c	c	c	a	c	c	c	1
	d	d	d	H	H	d	d	d	2
	e	e	e	e	e	H	e	e	1
	f	f	f	f	H	f	f	f	1
									(end of second cycle)
	H	a	a	a	a	H	a	a	2
	b	b	b	b	H	b	b	b	1
	c	d	H	c	c	c	c	c	2
	d	c	d	H	d	d	H	d	3
	e	e	e	e	H	e	e	e	1
	f	f	f	f	f	f	f	f	0
									(end of third cycle)

	cycle 1	ideal c.	cycle 2	ideal c.	cycle 3	ideal c.	Total	
D	4	0	10	0	9	0	23	0
C	45	49	32	42	33	42	110	133

D = number of deviations from the ideal section
C = number of concurrences (no deviation from the ideal section)

Are the deviations between the actual profiles and the ideal profile statistically significant, that is regionally relevant? This question can be answered easily by the methods of Chapter 11.

Next we investigate the deviation pattern as defined in Chapter 7 by comparing each profile with the ideal profile. For instance, we will obtain for the first profile the deviation pattern

abcHdef aHcdef Hbcdef
abcadef abcdef abcdef

0001000 010000 100000 (deviation pattern)

For the total of all seven profiles, the following runs of deviations will be obtained:

Length of run	Frequency	Number of deviations
1	17	17
2	1	2
3	0	0
4	1	4
	19	23

Mean length $L = (17 + 2 + 4)/19 = 1.211$
Mean length of corresponding random distribution
$= 133/(1 + 110) = 1.198$.

The difference is negligible, hence we cannot disprove the hypothesis that the deviations are distributed randomly and insignificant regionally. In this example, the actual mean length of the deviations, L, is close to unity. A value of L of 2 or larger would indicate that possibly deviations are correlated. Other approaches are given by Merriam (1964), Pearn (1964), and Carss and Neidell (1966).

In the last example, all cycles were recognizable easily. In many instances, however, this may not be the situation because the sequence of sedimentation deviates too much from any regular pattern. Then it may be helpful to investigate *transition frequencies* of different layers as indicated in the following example.

Example 8.2   A cycle from the Pennsylvanian System

The analysis is based on tables of Wanless (1939) which display a sedimentation pattern composed of approximately 350 beds in the Appalachian coal fields.

Suppose that there occurs the sequence abccabc, where a stands for sandstone, b for sandy shale, and c for conglomeratic sandstone. The transition from a to b and from b to c occurs twice. Hence the *transition frequency* of both a to b and b to c is 2. Also the transition frequency of a to b to c is 2, whereas the transition frequency of a to b to c to c is 1. Let us now consider the transition frequencies of the Pennsylvanian profile of the type I to II whereby the layer II is underlain directly by layer I:

	Absolute values							Relative values					
	I							I					
II	a	b	c	d	e	f	II	a	b	c	d	e	f
a	—	16	1	5	41	17	a	—	32	17	12	48	22
b	27	—	2	0	14	8	b	33	—	33	0	16	10
c	0	1	—	0	4	1	c	0	2	—	0	5	1
d	19	13	1	—	13	3	d	23	26	17	—	15	4
e	20	9	0	0	—	48	e	25	18	0	0	—	62
f	15	11	2	36	13	—	f	19	22	33	88	15	—
Sum	81	50	6	41	85	77	Sum %	100	100	100	100	99	99

a sandstone
b sandy shale
c conglomeratic sandstone
d clay, underclay
e shale, calcareous shale, carbonaceous shale
f coal

Such a scheme is termed a *transition matrix*. It shows that some combinations are particularly frequent. It further displays a high degree of asymmetry. For instance, the transition frequency f to e is large (62%) whereas the transition frequency e to f is small (15%). The e-columns of the transition matrices show that the layer e is followed preferentially by layer a. The transition frequencies e to b, e to d, and e to f are distinctly smaller and approximately equal. This suggests that the sedimentation of layer e does not create a preference for the subsequent sedimentation of say layer b rather than layer d or f. As a matter of fact, the significance of this hypothesis has to be tested (see Chapter 11).

A similar consideration of all columns reveals the following preferences:

a to b	(a underlain by b) frequency:	27	ss under s.sh
b to a		16	s.sh under ss
d to f		36	clay under coal
e to a		41	sh under ss
f to e		48	coal under sh

This suggests for instance the combination d to f, f to e, e to a, a to b (if the somewhat doubtful preference b to a is omitted). The combination corresponds to the profile

d	clay
f	coal
e	shale
a	sandstone
b	sandy shale

The smallest transition frequency is 27. Hence, the sequence dfeab could occur 27 times. Actually, however, it only occurs three times. On the other hand, the sequence dfea occurs eleven times and comprises 44 layers from a total of approximately 350. This shows that one must distinguish between the main preferences of two consecutive layers and the occurrence of larger sequences that may possibly form cycles.

As the next step we can consider the transition frequencies of three consecutive layers. It turns out that only one-half of all possible combinations are actually realized. The most frequent combinations are:

triad combinations:	fea	dfe	adf	ead	dfa	eae	bdf	edf
frequencies:	24	16	14	12	11	11	10	10

tetrad combinations:	dfea	edaf	aefe	feae	dfab	adfa	abdf
frequencies:	11	9	8	7	7	6	6

The following graph shows all tetrads which occur at least five times and the corresponding triads and dyads:

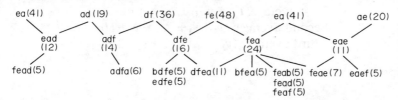

The observed transition frequencies may be interpreted as *transition probabilities* which determine the probability that a certain sediment is followed by a layer of a different petrography. An even deeper insight can be gained by applying the theory of Markov chains and Markov processes. The theory was developed by Markov when he investigated the transition frequencies of vowels to consonants in Pushkin's dramatical poem "Eugene Onegin". The reader who is interested in Markov chains and processes may consult the books of Gordon (1965) and Takacs (1964), a publication of Vistelius (1949), and the overviews

of Griffiths (1966) and Pettijohn, Potter and Siever (1965, p. 126–132).
This mathematical approach, however, is beyond the scope of this book.

Quantitative time series: Some fundamental concepts

Up to now we have considered sequences of layers that are ordered
qualitatively by the relation "deposited earlier—deposited later."
Chronological time or the thicknesses of the different beds did not enter
into the analysis. If, however, a time scale exists, or if the depth of a
petrographic unit is considered as an equivalent of physical time, then
the change of some property such as a sand/shale ratio may be studied
quantitatively as function of time or thickness. Thereby we must
distinguish among the *long-term trend* (i.e. the global change by and
large, *cyclical movements* with a long amplitude), *short-term regular
fluctuations* (such as *seasonal effects*) and *irregular movements* which do
not display any trend or cycle and which may be distributed randomly.
By definition, in statistics the term "seasonal" is not restricted to the
seasons such as spring and winter and obvious examples of so-called
seasonal effects are yearly temperature cycles and daily tidal cycles.

The long-term trend of a time series should be determined by the
method of least squares as explained in Chapter 13. This approach leads
to a unique result. There is, however, a simpler and faster way of
smoothing data. Suppose the time series is given by values y_1, y_2,
$y_3, \ldots, y_n, y_n+1, y_n+2, \ldots$ that refer to equidistant points of time or
equidistant depths. We select some integer n and calculate the *moving
averages of order n*:

$$\text{First average: } (y_1+y_2+ \ldots +y_n)/n$$
$$\text{Second average: } (y_2+y_3+ \ldots +y_n+1/n)$$
$$\text{Third average: } (y_3+y_4+ \ldots +y_n+2)/n$$

and so on.

Numerical example for n=3:

Sand/shale ratios:	0.2	0.6	0.1	0.5	0.3	0.7	0.2
Moving averages:		0.3	0.4	0.3	0.5	0.4	

because $(0.2+0.6+0.1)/3=0.3$
$(0.6+0.1+0.5)/3=0.4$ etc.

The sequence of moving averages represents an unidirectional long-term
trend if the smoothed values either increase or decrease during the span
of time being considered. There is a *stationary trend* if the smoothed
values roughly are constant. The example seems to indicate stationary

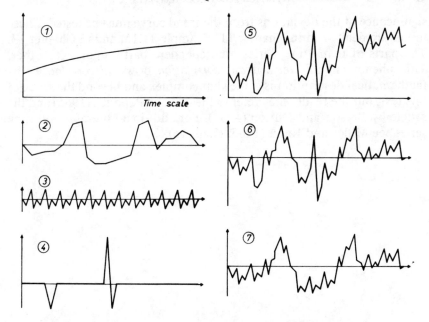

FIGURE 17. Deconvolution of time series. (1) Long-term trend. (2) Cyclical movements. (3) Strictly periodical seasonal effects. (4) Singular events, possibly indicating "catastrophe" (5) Actual time series. (6) Fluctuation curve (time series minus long-term trend). (7) Fluctuation curve, singular events omitted.

trend. To check this, a larger order n of the moving averages must be selected or the moving averages must be smoothed:

Moving averages:	0.3	0.4	0.3	0.5	0.4
Smoothed averages:	0.33	0.4	0.4		$(n=3)$

because $(0.3+0.4+0.3)/3=0.333$, etc. Obviously, the time series is too short to permit the establishment of a reliable long-term trend.

After having established the long-term trend, *trend elimination* is accomplished by subtracting the trend from the time series. The analysis of the resulting fluctuation curve may be rather tedious and somewhat ambiguous. First one should try to determine the period of any cyclic or quasicyclic movement. Then one should eliminate singular events. If this process can be performed successfully, the curve of seasonal fluctuations will be established (see Figure 17).

The decomposition can be done systematically by using *Fourier analysis*. This technique permits the decomposition of the fluctuation curve into "waves" of different lengths and amplitudes. The method will be considered in more detail in the chapter on geological maps.

Finally the reliability of the estimate of the long-term trend and the significance of the deviations from the trend curve must be tested. These topics are dealt with in Chapter 11, Example 11.15, and in Chapter 13. The parts of many time series are dependent partly upon each other. This phenomenon is termed *autocorrelation* or *serial correlation*. Its mathematical description is somewhat complex and beyond the scope of this text, but the reader may refer to the book by Papoulis (1965) and the article by Bryson and Dutton (1961). For additional literature on time series see Miller and Kahn (1965, chap. 15).

CHAPTER 9

Some Important Distribution Functions

9.1 Discrete Distributions

9.1.1 The Binomial Distribution (Bernoulli Distribution)

Suppose that a population consisting of m individuals is divided into two groups according to whether they have a certain attribute A. Such a division is said to be *dichotomous*. Let u individuals have the attribute A. Then $v = m - u$ individuals do not have the property A; $p = u/m$ is the relative frequency of individuals with property A, and $q = v/m$ is the relative frequency of individuals with the attribute not-A ($q = 1 - p$). When drawing a sample of n individuals, we will expect to draw np individuals with the property A and $n(1 - p)$ individuals with the property not-A. Actually, however, a sample that does not comprise the entire population may contain more or less than np individuals with attribute A. The sample may even contain no individual with property A or no individual with property not-A. In general, a random sample contains i individuals with property A where i can take any of the values 0, 1, 2, 3, ..., n. If we draw from a large population N samples of n individuals each, then we can expect that f_i samples contain i individuals with the property A, and

$$f_i = N C_{ni} p^i (1 - p)^{n-i}$$

The numbers C_{ni} can be calculated easily.

Table of the numbers C_{ni} for small values of n

n	i 0	1	2	3	4	5	6
2	1	2	1				
3	1	3	3	1			
4	1	4	6	4	1		
5	1	5	10	10	5	1	
6	1	6	15	20	15	6	1

The numbers in each row are calculated from the previous row by summation (Pascal). *Example*: Calculation of the last row of the table (from row 5): $1+5=6$, $5+10=15$, $10+10=20$, $10+5=15$, $5+1=6$. The first and the last value of every line are equal to unity.

The fractions f_i/N where

$$f_i/N = C_{ni}p^i(1-p)^{n-1} \qquad i=0, 1, 2, 3, \ldots, n$$

and

$$(f_0/N) + (f_1/N) + (f_2/N) + \ldots + (f_n/N) = 1$$

are termed the relative frequencies of the binomial (Bernoulli) distribution with parameters n (sample size) and p (expected relative frequency of the considered attribute A). It can be shown that the mean, variance, and third and fourth moment of the binomial distribution are

$$[\bar{x} = (f_1 + 2f_2 + 3f_3 + \ldots + nf_n)/N \text{ etc.}]:$$

$$\bar{x} = np$$
$$\sigma^2 = np(1-p)$$
$$\alpha_3 = (1-2p)/\sigma$$
$$\alpha_4 = 3 + (1/\sigma^2) - (6/n)$$

For $p=\frac{1}{2}$ the Bernoulli distribution is symmetric around the mean (cf. Figure 18A):

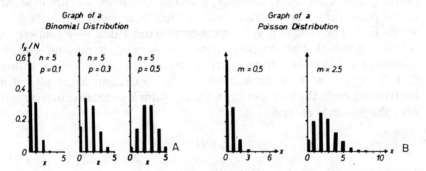

Graph of a
Binomial Distribution

Graph of a
Poisson Distribution

FIGURE 18. A. Binomial distributions. Mode moves to right as p increases; B. Poisson distributions (see section 9.1.3). Variable i where $i=0, 1, 2,\ldots$ is designated by the letter x.

Example 9.1 Comparing an actual distribution with the corresponding binomial distribution

From a sandstone $n=5$ grains are sampled at each of $N=60$ locations. (Actually, at least $n=30$ grains should be sampled at each location.)

Some grains belong to a certain class of minerals ("attribute A"). The following distribution is determined:

Number (i) of grains with attribute A:	0	1	2	3	4	5
Number (f_i) of locations:	12	22	8	6	8	4

Average number of grains with attribute A: 1.8

because $(0\times12+1\times22+2\times8+3\times6+4\times8+5\times4)/60=1.8$

Variance: $s^2=2.4$

If the mean 1.8 is considered as representative, the corresponding Bernoulli distribution has the parameters $n=5$ and $p=\bar{x}/n=1.8/5=0.36$. The corresponding variance $np(1-p)=1.2$ is only one-half as large as the variance of the actual distribution. The binomial distribution with $n=5$, $N=60$, and $p=0.36$ has the values:

i:	0	1	2	3	4	5
f_i:	6.44	18.12	20.38	11.47	3.22	0.36

It differs considerably from the actual distribution. This suggests that property A is not distributed at random in the sandstone. The hypothesis of nonrandom spatial distribution (i.e., of nonrandom clustering of certain minerals) must be tested (see Example 11.8).

9.1.2 The Multinomial Distribution

An immediate generalization of the binomial distribution is the multinomial distribution. It is concerned with the situation that the individuals of a population have one of several mutually exclusive attributes. For instance, a certain mineral may be either red or green or yellow. In such a situation, the multinomial distribution may be used to calculate the expected frequency that n_1 minerals are red, n_2 green, and n_3 yellow where $n=n_1+n_2+n_3$ is the sample size. The formulae for the multinomial distribution really are not difficult but somewhat cumbersome and will be omitted. In many instances it is sufficient to combine several attributes to a composite property thereby reducing the multinomial distribution to a binomial one. For example, one may compare the number of green minerals to the number of nongreen minerals.

9.1.3 The Poisson Distribution. The Frequency of Rare Events

The Poisson distribution is in a certain sense a particular limiting form of the binomial distribution. It can be used to calculate the frequency of a rare event or the frequency of a single property which does not occur often. The gist of the Poisson distribution most easily is explained by a geometrical analogy.

Let points or things such as minerals, rare fossils, or marks from rain drops each of which are "rare" events be distributed at random on a line, a surface, or in a three-dimensional geological body. Let k be the average number of points (or minerals, etc.) per unit length or unit surface or unit volume respectively. Further, let L be the length (area, volume) of that part of the geological body from which a sample will be drawn. Then we can expect in the sample $m=kL$ observations (marks of rain drops, fossils, etc.). Actually, however, more or less than m rare "events" may be present, possibly none at all. This is so because the observations (fossils, etc.) are not distributed regularly but randomly. When we draw a large number of samples, say N, each from a region of the same size L, then we can expect i rare events in f_i samples where $i=0, 1, 2, 3, \ldots, i$-max. If the region is of infinite size and if the rare events are distributed randomly,

$$f_i/N=(m/1)(m/2)(m/3)\ldots(m/i)\exp(-m) \qquad i=1, 2, 3, 4,\ldots$$

The expression $\exp(x)$ stands for the exponential function of the argument x. The expression $\exp(-m)$ is an abbreviation for $1/\exp(m)$. The relative frequencies f_i/N become negligibly small for large values of i. The formula indicates that the relative frequencies of the Poisson distribution are determined completely by one parameter, namely m, which is the average number of rare events in a region of specified size (mind that the Bernoulli distribution is determined by two parameters, the relative frequency p of the considered attribute, and the sample size n). It can be shown that

$$\bar{x}=m$$
$$\sigma^2=m$$
$$\alpha_3=1/\sqrt{m}$$
$$\alpha_4=3+(1/m)$$

Figure 18B shows two examples. Figure 19 displays a plot of the Poisson distribution for the important range $0.1 \leq m \leq 10$. The Bernoulli distribution can be replaced by the Poisson distribution with $m=np$ in the situation when np is smaller than 5 (cf. Example 9.4).

Example 9.2   Distribution of a rare event in a geological profile

A profile (120 ft) is divided into 120 sections of 1 ft each. The number of a certain rare event is determined for each section. On the average, there are two events per section. We compare the empirical count with

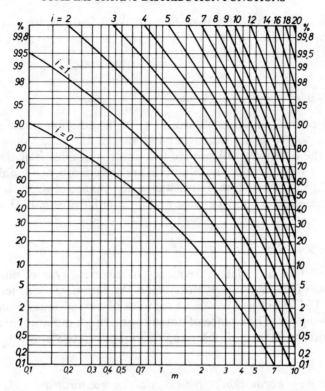

FIGURE 19. Poisson distribution. *Abscissa*: mean (logarithmic scale); *ordinate*: Cumulated frequencies F_i/N (%) for $i=0$, 1, 2, 3,...: $F_o/N=100f_o/N$, $F_1/N=(100f_o/N)+(100f_1/N)$, etc. Example: $m=2$: $F_o/N=14\%$, $F_1/N=40\%$, $F_2/N=68\%$. Hence 14% of N samples (N large) do not contain rare event, $40-14=26\%$ contain it once, $68-40=28\%$ contain it twice, and $100-68=32\%$ contain it at least three times.

the Poisson distribution for $m=2$. The match is close (F is the cumulative frequency):

i	f	Actual distribution $f/N(\%)$	$F/N(\%)$	Poisson, $m=2$ $F/N(\%)$
0	16	13.33	13.33	14
1	32	26.67	40.00	40
2	33	27.50	67.50	68
3	22	18.33	85.83	86
4	11	9.17	95.00	95
5	6	5.00	100.00	98.3
6				99.5
7				99.9
Sum	120	100	$\bar{x}=1.98$	

The comparison indicates that the rare event is distributed randomly. The theoretical probability of encountering the rare event more than five times in a section is very small (1.7%).

Example 9.3   Chance of survival

Suppose that a microorganism is given a radiation treatment. The radiation source emits quanta at random. If the lethal dose is $k+1$ quanta, the organism will survive if it absorbs at most k quanta. Thus, the chance of survival is the cumulated relative frequency for absorbing $i=0, 1, 2, 3, \ldots, k$ quanta. For instance, the chance of survival is 80% if the average dose is $m=10$ quanta and if the lethal dose is 13 (Figure 19).

Example 9.4   The distribution of mutation rates

Suppose that the relative frequency of a certain type of mutation is $p=0.004$. (This is not a typical value. Mutation rates differ widely; Rensch, 1947.) Let us investigate samples of $n=1000$ individuals each. Then $m=np=4$. The chance of not encountering a mutation in a sample is 2%, and the chance of observing one mutation is 8%.

Example 9.5   Formation of twin crystals in a saturated solution of K_2SO_4

By vaporizing potassium sulfate solutions at 20°C, Niggli obtained the following result (1948, p. 114):

No.	Crystals	Twins	
1	149	2	(1.3%)
2	49	0	(0.0%)
3	526	10	(1.9%)
4	186	2	(1.1%)
5	325	5	(1.5%)
Sum	1235	19	

Does the result of the second experiment with 49 crystals but no twin suggest that it might have been performed under conditions not in line with the other experiments? There are 19 twins from a total number of 1235 crystals. The average number m of twins from a sample of 49 crystals 19/1235 times 49, that is 0.8. According to Figure 19, for $m=0.8$ the chance is 45% that the rare event will not be encountered.

9.2 Continuous Distributions

9.2.1 The Normal Distribution

The *normal* or *Gaussian distribution* plays a key role in many parts of applied and mathematical statistics. The frequency distributions of many physical, biological, and geological phenomena are related closely to the normal distribution.

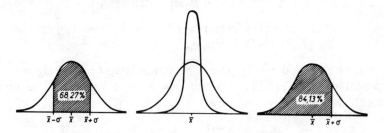

FIGURE 20. Normal (Gaussian) distribution. *Left-hand figure*: 68.27% of total frequency are in range from $\bar{x}-\sigma$ to $\bar{x}+\sigma$, 95.45% are in range $\bar{x}-2\sigma \dots \bar{x}+2\sigma$, and 99.73% are in range $\bar{x}-3\sigma \dots \bar{x}+3\sigma$. *Middle part*: Two normal distributions with mean $\bar{x}$ but different values of σ. Distribution is spread-out if variance σ^2 increases. Height of maximum at mode $\bar{x}$ is equal to $1/(\sigma\sqrt{\pi})$ and becomes smaller as σ increases. *Right-hand figure*: 84.13% of total frequency are in range between minus infinity and $\bar{x}+\sigma$.

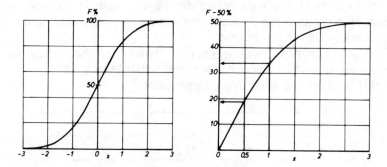

FIGURE 21. Cumulative frequency of normal distribution with mean zero and variance unity.

The normal (Gaussian) distribution is a limiting form of the binomial distribution and completely determined by two parameters, its mean $\bar{x}$ and its variance σ^2. The normal distribution is symmetric around the mean and represented by a bell-shaped curve (Figure 20). The mean, the median, the mode coincide: $\bar{x}=Q_2=M$. The variable x extends from minus infinity to plus infinity whereby the relative frequency of large values of x ("at the tails") is very small and can be ignored. Because variables representing natural phenomena are restricted to finite values, the normal distribution at the tails may be a poor approximation of an empirical distribution. In many instances, however, this does not really matter.

The normal distribution with mean $\bar{x}$ and variance σ^2 usually is referred to as $N(\bar{x}, \sigma^2)$. For instance, the symbol $N(0,1)$ refers to the Gaussian distribution with mean zero and variance unity (*standard form of the normal distribution*). The following formulae hold (the letter Y designates the ordinate of the Gaussian distribution):

$$N(0,1): Y=(1/\sqrt{2\pi})\exp(-x^2) \qquad -\infty<x<+\infty$$
$$N(\bar{x},\sigma^2): Y=(1/k)\exp(-z^2) \qquad -\infty<z<+\infty$$

where $k=\sigma\sqrt{2\pi}$ and $z=(x-\bar{x})/\sigma$. The symbol $\exp(-z^2)$ designates the reciprocal of $e=2.71828\ldots$ to the power of z squared. The higher moments are

$$\alpha_3=0 \qquad \alpha_4=3$$

Cumulative frequency F% versus x (also Figures 20, 21):

x	$\bar{x}-3\sigma$	$\bar{x}-2\sigma$	$\bar{x}-\sigma$	$\bar{x}$	$\bar{x}+\sigma$	$\bar{x}+2\sigma$	$\bar{x}+3\sigma$
F%	0.135	2.275	15.865	50	84.135	97.725	99.865%

The plot of $F\%$ versus x on *probability paper* results in a straight line. (Figure 19 represents the Poisson distribution on probability paper with logarithmic scale of the abscissa.)

The normal distribution closely approximates the Poisson distribution ($\bar{x}=\sigma^2=m$) if $m>10$ and the Bernoulli distribution ($\bar{x}=np$, $\sigma^2=np$ $(1-p)$) if both np and $n(1-p)$ are larger than 5:

Distribution	paragraph	$\bar{x}$	σ^2	
Binomial	9.1.1	np	$np(1-p)$	$np>5, n(1-p)>5$
Poisson	9.1.3	m	m	$m>10$
Iterations	7.2	$\bar{u}$	$\sigma^2(u)$	sample size>10

sample size=number of iterations

Example 9.6

The cumulative frequencies $F\%$ of an empirical distribution are plotted on probability paper. The plot results approximately in a straight line from which the following values are read: $F=50\%$, $x=13$; $F=84\%$, $x=17$. Calculate $\bar{x}$ and σ. *Answer: $\bar{x}=13$, $\bar{x}+\sigma=17$, $\sigma=4$.*

Example 9.7

Calculate for $N(0,1)$ the relative frequencies of the classes $0.5 \leq x \leq 1$ and $-0.5 \leq x \leq 0.3$.

Answer: $34.13 - 19.15 = 14.98\%$ and $19.15 + 11.79 = 30.94\%$ (use the table of the normal distribution in the Appendix).

Example 9.8

Calculate for $N(1,4)$ the relative frequency of the class $0 \leq x \leq 3$. *Answer:* $x = 0$: $z = (0-1)/2 = -0.5$. If $x = 3$ then $z = (3-1)/2 = 1$ (mind that $\sigma = \sqrt{4} = 2$). Hence the relative frequency is $19.15 + 34.13 = 53.28\%$.

Example 9.9

Calculate the ordinate Y of $N(1,4)$ at $x = 3$.

Answer: $z = (3-1)/2 = 1$.   Hence $\sigma Y = 0.240$ and $Y = 0.242/2 = 0.121$

Example 9.10

Calculate the value of the binomial distribution for $p = 0.4$ and $n = 20$ (i.e., $\bar{x} = 8$ and $\sigma^2 = 4.8$) at $x = 3$. The normal distribution can be used because $np = 8$ and $n(1-p) = 12$ are both larger than 5. The binomial distribution is defined for the discrete values $x = 0, 1, 2, 3, \ldots, 20$. Thus, the value of the binomial distribution at $x = 3$ corresponds closely to the class $2.5 \leq x \leq 3.5$ of the normal distribution $N(8, 4.8)$.

Numerous parameters x such as grain sizes, rock permeabilities, and layer thicknesses are distributed approximately normally if the parameter $y = \log x$ rather than x is considered as the random variable (*log normal distribution of x*). In such a situation one may use the normal distribution of y with the mean $\bar{y}$ and the variance σ^2 of y.

9.3   Systems of Distribution Functions

Thus far we have dealt with a few special distributions which arise rather frequently. Several attempts have been made to develop a general system of distributions which can describe or at least closely approximate various types of different families of distributions of a random variable. The *system of Pearson* and the *Gram-Charlier series* should be mentioned only. Historically, all distributions mentioned so far were derived from the consideration of physical situations such as the tossing of coins or the distribution of observational and experimental errors. The system of Pearson, on the other hand, is derived from purely formal considerations as the solution of a certain differential equation. There are twelve types identified by Roman numbers. For instance, type VII is the normal distribution and type III constitutes the Chi-square distribution that shall be considered in Chapter 11.3. The Gram-Charlier series are more general. In fact, they may be used to describe almost any empirical distribution by a mathematical formula that involves the moments of the empirical frequency distribution (see Wilks, 1946, p. 76).

9.4 Interpolation Formulae

Several distribution functions used in the literature simply are interpolation formulae which constitute a compromise between simplicity of the formula and accuracy of the description. As an example: The description of grain-size distributions by the *RRS-formula* (*Rosin's law*) can be mentioned (Gebelein, 1956; Kittleman, 1964; Bennett, 1936; and Theimer, 1952). The genetic interpretation of empirical distributions is a delicate and possibly precarious undertaking unless there exists a sound quantitative physical or biological model. For instance, under certain assumptions, the log normal distribution may be derived from models of crystallization or mechanical fracturing. On the other hand, the log normal distribution seems to be an interpolation formula with respect to grain-size distributions originating from transport processes in river etc. (cf. Kottler, 1950; Kolmogoroff, 1941; and Epstein, 1947.)

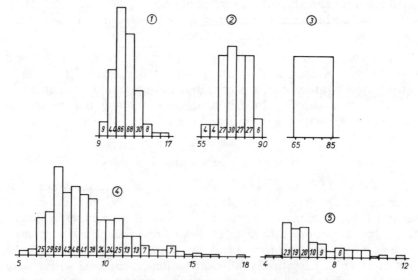

FIGURE 22. Variability of certain quantitative properties of *Ceratites* according to Wenger (1957). For details see text. Subfigures 1, 2, 4, and 5 correspond to figures 22, 13, 25, and 14 respectively of Wenger's monograph.

Example 9.11   Variability of Ceratites

In a monograph, Wenger (1957) represented a large number of measurements performed on *Ceratites* and discussed the taxonomic and phylogenetic consequences. Figure 22 shows some of his results. Figure 22(1) indicates the variation of the number of lateral knots of the last whorl. This distribution represents a general empirical type; a bell-shaped curve with insignificant skewness. It can be approximated by a normal distribution with the same mean and variance. Figure 22(2) shows the variation of the cross section of the last septum. It corresponds closely to the constant-frequency distribution of Figure 22(3). Figures 22(4) and 22(5) show the variability of the diameter of the phragmacons. They represent approximately bell-shaped curves with a rather long-drawn out right-hand-tail and a significant skewness. In most situations, the diameter of the phragmacon is small; occasionally, however, it is rather large. In many instances, the distributions of positive quantities such as lengths are skew-symmetric if the modes are relatively near to zero because large positive values may occur whereas negative ones cannot.

CHAPTER 10

Estimating Population Parameters, Confidence Limits. The Student-t Test

Suppose we draw different samples from the same universe (population). As a consequence of the variability of the data, the parameters of the samples such as mean, mode, variance, etc., will in general differ from sample to sample, and no sample parameter may coincide with the corresponding parameter of the universe. For instance, the sample means may be 14, 27, 7, 18, 13, 48, 15,..., whereas the population mean is equal to 19.

If, however, the sample size is very large in terms of the number of different samples or if the variance of the population is small, then the sample may be considered as nearly representative of the universe. For the sample yields the correct population parameters if the sample comprises the whole universe or if the universe's variance is nil.

Irrespective of the properties of the universe and the size of the sample, the population mean $\bar{x}_p$ always will be located in some interval of numbers. More specifically, $\bar{x}_p$ will be located between the limits $\bar{x}_s - L$ and $\bar{x}_s + L$ where L is some number and $\bar{x}_s$ is the sample mean. Unfortunately, the number L is either unknown or so large that the bounds for $\bar{x}_p$ are meaningless. One, however, may calculate a different number, l—one that may be small relatively and guarantees that $\bar{x}_p$ lies between $\bar{x}_s - l$ and $\bar{x}_s + l$ in a specified percentage of situation, say 99% of the time. Thus for ninety-nine drawn samples the population mean will be located in the stated interval but will be out of range for one sample that by chance represents a collection composed of data mainly from one of the tails of the total distribution. The sample mean $\bar{x}_s$ is termed an *estimate* of the population mean $\bar{x}_p$ whereby $\bar{x}_p$ is located between the *confidence limits* $\bar{x}_s - l$ and $\bar{x}_s + l$ with the *confidence level* 99%. In other words, the chance is 1% that the population mean is located *outside* the interval with the limits $\bar{x}_s - l$ and $\bar{x}_s + l$. Hence in technical terms "the result is *significant at the 1% level*."

Obviously, the number 1 is small if the sample size N is large (N is the number of observed values of the sample), or if the variance σ^2 of the

58

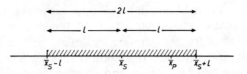

FIGURE 23. Confidence limits (also termed fiducial limits). For details see text.

universe is small, or if the confidence level is small. If N at least is equal to 30, the number l can be estimated easily from N and the standard deviation s of the sample by the following formulae:

Parameter	l (95%)	l (99%)
$\bar{x}$	$1.96 \cdot s/\sqrt{N-1}$	$2.58 \cdot s/\sqrt{N-1}$
Q_2	$2.46 \cdot s/\sqrt{N-1}$	$3.23 \cdot s/\sqrt{N-1}$
σ	$1.39 \cdot s/\sqrt{N-1}$	$1.82 \cdot s/\sqrt{N-1}$
σ^2	$2.77 \cdot s/\sqrt{N-1}$	$3.65 \cdot s/\sqrt{N-1}$
Q_1, Q_3	$2.67 \cdot s/\sqrt{N-1}$	$3.52 \cdot s/\sqrt{N-1}$
D_1, D_9	$3.35 \cdot s/\sqrt{N-1}$	$4.41 \cdot s/\sqrt{N-1}$
D_2, D_8	$2.80 \cdot s/\sqrt{N-1}$	$3.69 \cdot s/\sqrt{N-1}$
D_3, D_7	$2.58 \cdot s/\sqrt{N-1}$	$3.40 \cdot s/\sqrt{N-1}$
D_4, D_6	$2.49 \cdot s/\sqrt{N-1}$	$3.27 \cdot s/\sqrt{N-1}$

Remarks

(1) Strictly speaking, the universe should be distributed normally or at least approximately normal.

(2) To estimate the confidence limits of σ and σ^2, the sample size N should be at least 100.

(3) For log normal distributions, $\bar{x}$, Q_2, etc., refer to the logarithms of the data.

Example: Let Q_2 be the median of a sample of size $N=82$ drawn from a binomial distribution. Then it is a fair estimate to locate the population median in the interval with the limits $Q_2-(3.23s/9)$ and $Q_2+(3.23s/9)$. This estimate is significant at the 1% level.

From the table it can be determined that the mean has a smaller confidence interval than the median ("$\bar{x}$ is more *efficient* than Q_2") and that the median has a smaller confidence interval than the other quartiles.

The confidence interval for the population mean can even be determined if the sample size N is small (N may be as small as 2). The corresponding formula developed for this situation originated from W. S. Gosset who published under the pseudonym "Student". He introduced a parameter labeled t that depends on a variable v termed "degrees of freedom" in mathematical statistics. In the Appendix Table III, t versus v is tabulated for the confidence levels 95% and 99%. In the following formula, "the degrees of freedom" are $v=N-1$.

If $\bar{x}$ and s are the sample's mean and standard deviation respectively the population mean is in the range

$$\bar{x} \pm t \cdot s/\sqrt{N-1}$$

It should be kept in mind that there are no methods to estimate population parameters if the sample size N is unknown or not defined. N is always an integer such as the number of mineral grains of a sample or the number of excavated skulls of a certain type of animal. If the sample is represented by percentages such as weight percentages or other relative numbers, then N is not defined and neither the stated nor the following methods can be applied. In such a situation the following procedure may be used: Draw N samples (say $N=17$) and determine the median (or mean, etc.) of each sample. There results a single new sample composed of $N=17$ numbers with a certain mean $\bar{x}$ and standard deviation s. We can consequently estimate the confidence interval of a population that consists of all medians (or means, etc.) of the original universe.

Example 10.1   Thickness variation of an oil reservoir

An oil reservoir is being produced from $N=17$ wells that seem to include all different parts of the pool and are distributed according to a regular pattern. All wells fully penetrate the pay zone, whose mean thickness as calculated from well logs is $\bar{x}=20$ feet. The standard deviation is $s=0.8$ feet. The degrees of freedom are $17-1=16$. The corresponding value of t at the confidence level 95% is $t=2.12$ (Appendix, Table III). The confidence interval is $20 \pm 2.12 \times 0.8/4 = 20 \pm 0.4$. Hence, the average thickness of the pay zone is in the range between 19.6 and 20.4 feet. This result is significant at the 5% level, that is the chance is 5% that the thickness is smaller than 19.6 feet or larger than 20.4 feet.

Example 10.2   Confidence rectangles

Suppose there is a sample of N points that are located on a map. Let this spatial distribution of N points reflect the actual distribution of some geologic or petrographic phenomenon. Each sample point is fixed by two coordinates such as its latitude and longitude. Then we may calculate the two coordinate means and the corresponding confidence limits. The two means define "the point of center of gravity" of the sample points. The confidence limits define a rectangle that contains "the point of center of gravity" of the total distribution at a certain significant level. In the situation where there are two samples, the corresponding rectangles may overlap such as in Figure 24. This may indicate that the deviation of the two points of gravity is not significant. For details see Example 11.4

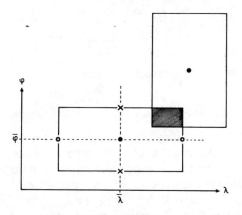

FIGURE 24. Two confidence rectangles that are overlapping.

Example 10.3   Enumeration of mineral components in a thin section of a rock by the point-counting method

The percentage volume of a mineral labeled A is approximately equal to the percentage area of A in a representative cross section of the rock; see Glagolev (1934) and Chayes (1956). This assumption tentatively is made when laying a grid of n points over a thin section of the rock and counting the numbers n_A of grid points that are located on A. Then the ratio n_A/n is considered as the relative volume of the mineral A. The accuracy of the result depends on several factors such as the number and distribution of grid points, the inhomogeneity and grain-size distribution of the rock, and the total area investigated. *Numerical example*: Figure 15B (Chapter 7, Example 7.3) is divided into four quadrants of thirty-six squares each. A signifies the dark component. Grid points that are on more than one quadrant are only counted once.* There results a sample consisting of $N=4$ values of the ratio n_A/n:

*Such grid points are located on the common boundaries of quadrants.

Quadrant	n_A	n	n_A/n %
upper left	14	36	39
upper right	17	42	40.5
lower left	16	42	38
lower right	17.5	49	36.5

$$\bar{x} = 38.5\%$$
$$s^2 = 2.1$$
$$t = 3.18 \ (95\%)$$
$$\text{Range: } 38.5 \pm 2.7$$

Actually, 38% of the total area is composed of the dark component.

Example 10.4   How to estimate the necessary sample size N

If the sample size N is large, the confidence interval is rather small and the mean of the sample size is close to the mean of the universe. On the other hand, it may be too time-consuming or even impossible to draw a large sample. Hence the question arises, "can we estimate a sample size N that is relatively small but large enough to yield a sufficiently representative mean?" This question may be answered approximately by using the following empirical device.

As a rule of thumb, the product "ts" that is "t-value times standard deviation of the sample," only changes moderately if N is larger than say 25. Thus, the confidence range roughly is halved if a sample of size $N=30$ is replaced by a sample of size $N=100$ ($\sqrt{100}=10$, $\sqrt{30}\approx 5$). Further, the confidence range for the sample size $N=250$ is about one-third of the confidence range for $N=25$ ($\sqrt{250}\approx 15$, $\sqrt{25}=5$). It should be kept in mind that these figures are rough estimates giving a hint of the necessary sample size after having drawn a sample of size $N=25$ to $N=30$.

Example 10.5   The delimitation of taxonomic units

The following refers to rhinoceros species from the European Pleistocene. For a long time, the existence of the species *Dicerorhinus hemitoechus* has been questioned; corresponding fossil remains have been assigned to *D. kirchbergensis*. A recent unpublished investigation of K. D. Adam revealed distinct differences of the molars of the upper jaws. The significance of these differences shall be considered in this example for the premolars P3 and P4. With respect to *D. hemitoechus*, the teeth are from the middle Pleistocene of the Heppenloch cave at the State of

Baden-Württemberg, West Germany. The teeth of *D. kirchbergensis* are from the upper and middle Pleistocene of different parts of southwestern Germany.

P3

	Mean		Variance		Number of molars	
	D.h.	*D.k.*	*D.h.*	*D.k.*	*D.h.*	*D.k.*
Length (mm)	37	41	0.72	5	6	9
Breadth	49	60	1.37	7	6	9

P4

	Mean		Variance		Number of molars	
	D.h.	*D.k.*	*D.h.*	*D.k.*	*D.h.*	*D.k.*
Length (mm)	41	48	0.34	7.3	5	16
Breadth	57	71	5.8	7.4	5	16

Breadths are measured from the front.

Figure 25 shows a plot of length versus breadth and the corresponding "confidence rectangles." The premolars P3 *D.h.* differ significantly from

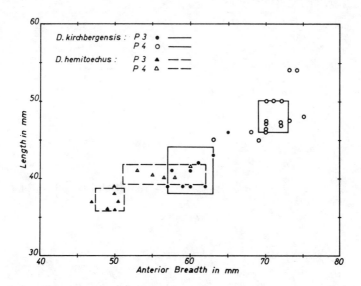

FIGURE 25. Length-breadth diagram of premolars P3 and P4 of two species of Pleistocene rhinoceros from southwestern Germany.

the premolars P3 *D.k.* The same holds for the premolars P4. The considered parameters of the premolars P3 *D.k.* and P4 *D.h.* overlap.

This is a typical example of delimiting fossil biological forms by statistical methods. Its advantages and disadvantages are discussed by Sylvester-Bradley (1958).

CHAPTER 11

Significance Tests for Means, Variances, Frequency Distributions, and Sequential Arrangements

In the last example of the previous chapter we compared the means of several samples of fossil remains to determine whether the samples belong to the same biological species. In this chapter the methods of comparing samples will be systematized, generalized, and put on a firmer basis. The fundamental question of this chapter is, "suppose, there are two samples—do then these samples belong to the same universe (population)?"

We shall try to answer this question for any specific situation by performing one or several *tests*. In the parlance of the statistician, a test is a procedure for accepting or rejecting a *statistical hypothesis*. In general, a statistical hypothesis is any statement relating one or several samples to some universe or some universes. For instance, the hypothesis may be that the mean $\bar{x}$ of a sample drawn from a normal universe with mean a is equal to a: $\bar{x} = a$. As a general rule, a test is set up with the hope of rejecting the hypothesis. For this reason, the hypothesis usually is termed a *null hypothesis H_o*. Thus, one may want to confirm the suspicion that two populations have different means. Then one takes as H_o the hypothesis that the means are equal. If H_o is rejected by the test, then the suspicion is confirmed on the basis of the test used.

It is up to the investigator to be more or less exacting concerning the confidence level required to admit that the test has demonstrated a significant result. It is conventional among many workers either to select the 1% or the 5% significance level, that is a confidence level of 99% or 95% respectively. Other workers in the field prefer a more conservative level of significance, as for instance a confidence level of 99.9%, that is a significance level of 0.1%. Yet other workers adopt the following rules. (1) A result is not significant if the confidence level is smaller than 95%. (2) A result is significant if the confidence level is at least 99%.

(3) Additional tests are necessary if the confidence level is between 95% and 99%.

Strictly speaking, there is no possibility whatsoever of determining confidence levels that are reliable in all situations. A good judgement can come only from experience in the particular field in which the problem occurs. Yet how can a geoscientist get experience in how to select reliable confidence levels? There is only one answer: by comparing the results gathered from a few small samples with the results obtained from a large number of large samples. In nearly all situations, however, the geologist has neither the time nor the opportunity to collect a large number of large samples and must be content to draw and to analyze a small number of small samples. Further, account must be taken that most universes and fossil populations that occur in the geosciences display a high degree of variability. In general, the fluctuations among different samples from the same universe will be rather large. Thus a 99.9% confidence level cannot be recommended but a level of 95% where the fluctuations are known to be large or moderately large, and the 99% level otherwise.

The reader should differentiate carefully between tests that are concerned with single parameters and tests that compare complete frequency distributions. A series of tests may result in the statement that the means and the variances of the samples do not differ significantly. This statement does not imply necessarily that the samples originate from the same universe. They should originate from the same universe if the latter is determined completely by its mean and its variance such as are the normal, binomial, and Poisson distributions, otherwise the samples may originate from different populations (Figure 26).

FIGURE 26. Graphs of different distributions that share same means and same variances.

11.1   The Variance Ratio Test (Snedecor's *F*-test)

Snedecor's *F*-test permits testing the null hypothesis that the variances of the parent populations of two samples are identical (or at least nearly equal). The test rests on the assumption that the universes from which the samples were drawn are normal distributions or distributed approximately normal. We calculate the unbiased estimates $\hat{s}^2$ of the sample variances s^2 (see Chapter 5) and compute the ratio

$$F_{ratio} = \frac{\text{greater value of the estimates}}{\text{lesser value of the estimates}},$$

This ratio is compared to the corresponding F-value of Table IV in the Appendix by using as degrees of freedom the sample size minus unity. If the ratio is larger than F, then the variances differ significantly. In this situation the parent populations of the samples are different. If, however, the variances do not differ significantly, the samples were either drawn from the same population or from different universes with the same variance.

We also may test the hypothesis that the unbiased estimate $\hat{s}^2$ of a single sample does not differ significantly from the variance σ^2 of a known or specified universe U.

Example 11.1   Numerical example

From Example 9.1 the following data were obtained (see Chapter 9, binomial distribution).

Empirical distribution: $N=60$ $s^2=2.36$ $\hat{s}^2=2.36\times60/59=2.40$
Binomial universe: $N=\infty$ $\sigma^2=1.15$
Ratio$=2.40/1.15=2.1$
Degrees of freedom for the greater variance: 59
Degrees of freedom for the lesser variance: ∞
Snedecor's F-value: between 1 and 1.4

Conclusion: the difference of the variance is significant.

11.2   The Student-t Test for Comparing Means

Use Table III of the Appendix. The parent populations should have normal distributions; however, investigations have shown that the test also is reliable if the universe deviates from the normal distribution. The sizes of the samples may be small or large, but the variances of the compared samples or populations should not differ significantly (check by using the variance ratio test). If they do, apply the procedures labeled as cases 4 and 5. The Greek letter v designates the degrees of freedom; the triple $(N, \bar{x}, s^2)$ stands for "sample size N, sample mean $\bar{x}$, sample variance s^2." ABS(a) is the absolute value of a, SQR(b) the square root of b and s the square root of s^2.

Case 1: Comparison of a sample mean $\bar{x}$ with the mean $\bar{X}$ of a specified very large or infinite universe. Calculate

$$v = N-1$$
$$A = \text{ABS}(\bar{X} - \bar{x})$$
$$B = \text{SQR}(v)$$
$$t = AB/s$$
(A times B divided by s)

Case 2: Comparison of $(\bar{x}_1, N, s_1)$ and $(\bar{x}_2, N, s_2)$. Calculate

$$v = 2N-2$$
$$A = \text{ABS}(\bar{x}_1 - \bar{x}_2)$$
$$B = (N-1)/(s_1^2 + s_2^2)$$
$$C = \text{SQR}(B)$$
$$t = AC$$

Case 3: Comparison of $(\bar{x}_1, N_1, s_1)$ and $(\bar{x}_2, N_2, s_2)$. Calculate

$$v = N_1 + N_2 - 2$$
$$s_0^2 = (N_1 s_1^2 + N_2 s_2^2)/v$$
$$A = \text{ABS}(\bar{x}_1 - \bar{x}_2)$$
$$B = N_1 N_2/(N_1 + N_2)$$
$$C = \text{SQR}(B)$$
$$t = AC/s_0$$

The compared means of cases 1, 2 and 3 differ significantly if the computed t-value is larger than the corresponding t-value of Table III.

Case 4: Comparison of a sample mean with the mean of a specified universe. The variances differ significantly. Calculate the confidence interval of the sample mean. The means differ significantly if the mean of the universe is not located within the confidence interval of the sample mean (compare with the previous chapter).

Case 5: Comparison of two sample means. The variances of the samples differ significantly. Calculate the confidence intervals of the means. If they do not overlap, then the means differ significantly.

Example 11.2   Testing nonnumerical and relative frequency distributions

Every significance test requires the sample size N, that is the number of items that constitute the sample. Unfortunately, sample sizes are not known if they are not stated and if all frequencies are relative values such as percentages. Another difficulty arises if the classes of the distribution are defined by mineral names or by nonnumerical properties such as colors. Then mean and variance do not exist and cannot be calculated.

Nevertheless, these cases can be treated if several samples are drawn from the same universe.

Case A: Each sample constitutes a relative frequency distribution. There is a total of N samples.

Calculate some parameter p such as the median or the mean of each relative frequency distribution. There result N numbers p_1, p_2, p_3, ... which constitute a sample of size N. This new sample may be compared with a corresponding second sample.

Case B: Each sample constitutes a distribution with nonnumerical classes such as rock minerals. There is a total of N samples.

For each class (such as quartz) there is a sample of size N whose mean and variance can be calculated and compared with the corresponding values of another set of samples.

Example: The composition of a rock is investigated at five different locations. The percentages by volume of quartz are 18.9, 21.3, 19.0, 20.2, 19.7. This is a sample of size $N = 5$.

A different approach to distributions with nonnumerical classes is shown in Chapter 12.

Example 11.3   The delimitation of different facies types which cannot be distinguished easily visually

Enumerate different rock properties which may be used to define several facies and describe each facies by means of average values of the different properties. Two facies types may be termed different if they differ significantly for at least one relevant property.

Example 11.4   Testing the significance of the distance between two points

Let a point P be defined by the mean coordinate values of the points of some point set (Example 10.2). In general, the two mean points of two different point sets will not coincide. The significance of noncoincidence may be tested by considering each coordinate separately. The distances between the points of the two sets can be calculated and the hypothesis tested that the mean distance is zero.

Example: Points in three-dimensional space are determined by the Cartesian coordinates x, y, z. Let these coordinates be

First point set							Second point set				
x	y	z					x	y	z		
1	4	3					0	0	0		
2	3	3					1	0	0		
2	5	4					1	0	1		
							1	1	1		
$N_1=3$							$N_2=4$				(sample sizes)
x-samples:			1,	2,	2		0,	1,	1,	1	
y-samples:			4,	3,	5		0,	0,	0,	1	
z-samples:			3,	3,	4		0,	0,	1,	1	

Compare first the x-samples by using the formulae of case 3, then compare the y-samples and finally the z-samples. Is there at least one significant difference? As an alternative, each of the three points of the first point set may be compared with each point of the second point set. There result $3 \times 4 = 12$ distances among the points of the two sets. These distances constitute a sample of size $N=12$ and may be compared to a hypothetical distribution with mean zero by using the formulae of case 1.

Any reader who wants a more rigorous approach to the problem of comparing simultaneously the means of several distributions should consult some advanced text on statistics, which refers to Hotelling's T^2 test.

Example 11.5　Mutations of the coral Calceola sandalina: *the significance of bimodality*

The coral *Calceola sandalina* (*C.s.*) has a shape resembling a shoe. The angle of the tip of the "shoe" can be measured rather accurately. Lotze (1928) investigated the tip angles of *C.s.* from three Devonian subformations. He determined a broad form (*C.s. sandalina*) in the higher Sparganophyllum limestone, the narrower form *C.s. alta* in the Upper Newberry layer and a bimodal sample in the Lower Sparganophyllum limestone which is located between the higher Sparganophyllum limestone and the Upper Newberry layer (Figure 27). The last sample represents possibly a mixture of *C.s. sandalina* and *C.s. alta*. Unfortunately, the last sample consists only of fifty-three specimens. Hence, the bimodality might not be representative of the population whose distribution might be monomodal. In this situation the evolutionary condition would be different.

To examine all possibilities, we compare first the actual bimodal distribution with a corresponding monomodal one which also is composed of fifty-three corals:

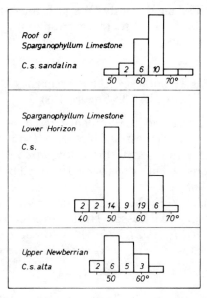

FIGURE 27. Variation of tip angles of coral *Calceola sandalina*.

Angles (class midpoints)	40°	45°	50°	55°	60°	65°	70°
Number of corals (actual)	2	2	14	9	19	6	1
Number of corals (regrouped)	2	2	11	19	12	6	1

The comparison of the two distributions by the Chi-square test explained in the next paragraph does not result in a clear answer: a small shift of the single mode of the regrouped sample results in a shift from "significant" to "not significant." Let us therefore try a different approach by separating the bimodal distribution into two monomodal ones that are compared by applying the formulae of case 3:

Angles	40°	45°	50°	55°	55°	60°	65°	70°
Number of corals	2	2	14	4	5	19	6	1
Sample size			$N=22$			$N=31$		

$$\bar{x}_1 = 49.6$$
$$\bar{x}_2 = 60.5$$
$$S_1^2 = 15.7$$
$$S_2^2 = 11.9$$
$$S_0^2 = 14.0$$
$$t = 10.5$$
$$t_{99=} = 2.67$$
$$S_0 = 3.74$$
$$v = 22 + 31 - 2 = 51$$
$$t > t_{99}\%$$

The two means differ significantly. We also may say the same of the modes which are both close to the corresponding means. For more details see Marsal (1949).

11.3   Testing Hypotheses About Frequency Distributions: The Chi-Square Test

The variance ratio test and the Student-t test refer to single although important parameters. To test a hypothesis about a whole frequency distribution, we have to resort to the *Chi-square test.* (A special test for circular distributions of Tukey has been described by Rusnak (1957); see also Middleton (1965).)

Let E_1, E_2, E_3, ..., E_k be k "events" which are interpreted as numerical or nonnumerical classes. The *observed* corresponding absolute frequencies are f_1, f_2, f_3, ..., f_k. We want to compare this to *expected* absolute frequencies v_1, v_2, v_3, ..., v_k with the same sample size N:

$$N = f_1 + f_2 + f_3 + \ldots + f_k = v_1 + v_2 + v_3 + \ldots + v_k.$$

The number of classes, k, should not be larger than twenty, and no expected frequency should be smaller than five. If possible, the smallest expected frequency should be ten or larger. This usually can be achieved by combining two or more classes. We calculate a number χ_s^2 by using the following scheme:

Event	E_1	E_2	...	E_k
Observed frequencies	f_1	f_2		f_k
Expected frequencies	v_1	v_2		v_k
Positive difference	$\lvert f_1 - v_1 \rvert = d_1$	$\lvert f_2 - v_2 \rvert = d_2$		$\lvert f_k - v_k \rvert = d_k$

$$d_1^2/v_1 = D_1 \quad d_2^2/v_2 = D_2 \ldots d_k^2/v_k = D_k$$
$$\chi_s^2 = D_1 + D_2 + D_3 + \ldots + D_k$$

If χ_s^2 is large, the expected distribution differs considerably from the observed one. If χ_s^2 is zero, the two distributions are identical. To be more specific, χ_s^2 is compared with the values of $\chi_{99}^2\%$ ("1% level") and $\chi_{95}^2\%$ ("5% level") of Table V in the Appendix. The compared distributions differ significantly if $\chi_s^2 > \chi_{99}^2\%$. The hypothesis that the two distributions do not differ significantly cannot be rejected if $\chi_s^2 < \chi_{95}^2\%$. In all other situations the significance is open to suspicion. The lines drawn by these rules are conventional but in general reliable. A somewhat different approach may be used (see example).

The actual χ^2 value of Table V depends on the so-called "degrees of freedom," v. In many instances, v is equal to $k-1$, that is the number of classes minus one. This holds if the distribution, v_1, v_2, v_3, ..., v_k is an empirical distribution. More generally, it holds if the distribution $f_1, f_2, f_3, \ldots, f_k$ is not required for determining the distribution $v_1, v_2, v_3, \ldots, v_k$. *Example:* we test the hypothesis that the f-distribution does not differ significantly from equipartition, that is we compare the f-distribution with the v-distribution $v_1 = v_2 = v_3 = \ldots = v_k$.

It also may happen that h parameters of the f-distribution are required to determine the v-distribution. In this situation the number of degrees of freedom, v, is equal to $k - 1 - h$.

Example: we test the hypothesis that the f-distribution is a sample drawn from a Poisson distribution. The latter is determined completely by one parameter, namely the mean m. Hence, we calculate the mean m of the f-distribution and determine the v-distribution as a Poisson distribution with mean m. In this situation we use one parameter derived from the f-distribution. Hence, $h = 1$ and $v = k - 1 - 1 = k - 2$.

If $v = 1$, or if most values of the v-distribution are smaller than 10, the *corrections of Yates* should be used to calculate χ_s^2:

$$D_1 = (d_1 - 0.5)^2 / v_1 \qquad D_2 = (d_2 - 0.5)^2 / v_2 \text{ etc.}$$

The Chi-square test can be generalized easily by introducing the concept of *contingency table*. In fact, this is done in most applications. A contingency table is a rectangle subdivided into boxes. Each box is filled with a number which represents an absolute frequency.

The smallest possible contingency table is composed of $2 \times 2 = 4$ boxes. We subdivide a square into four boxes filled with the absolute frequencies f_1, f_2, f_3, and f_4. For instance, these frequencies may indicate the following:

f_1 members of the sample have property A and property B,
f_2 members of the sample have property not-A and property B,
f_3 members of the sample have property A and property not-B,
f_4 members of the sample have property not-A and property not-B,

(sample size $N = f_1 + f_2 + f_3 + f_4$):

	A	NA
B	f_1	f_2
NB	f_3	f_4

Thereby property A may be the color "green" and property B a certain grain-size class. Slightly larger is a contingency table of $2 \times 3 = 6$ boxes. We subdivide a rectangle into 2×3 boxes. The corresponding frequencies refer to any combination of one of the properties A, B, C with one of the properties S or T:

	A	B	C
S	f_1	f_2	f_3
T	f_4	f_5	f_6

For instance, f_5 items of the sample have property B as well as property T. The properties A, B, and C may stand for grain-size classes or anything else, the letter S may stand for the mineral quartz, and the letter T may stand for the mineral feldspar. The examples show that the "properties" or "events' may be numerical or nonnumerical classes. Thus, we may set-up a 6×12 contingency table representing six different types of minerals which may occur in twelve different shades of color.

The actual calculation of χ_s^2 for a contingency table is simple. There must be one table for the f-distribution and a second table of the same type for the v-distribution. The value of each box of the f-distribution is compared to the value of the corresponding box of the v-distribution. For each pair of boxes $D = d^2/v$ is calculated. The sum of all D-values is equal to χ_s^2. Now let each contingency table be composed of m times n boxes, and let h be the number of parameters needed to "estimate", that is to calculate, the v-distribution. Then the number of degrees of freedom, v, is

$$v = (m-1)(n-1) - h$$

For a 2×2-table, $m = n = 2$ and $v = 1 - h$. However, the smallest possible value of v is 1. Hence, h must be zero: for a contingency table with 2×2 boxes, no parameter of the v-distribution can be estimated from the f-distribution. The v-distribution must be constructed without referring to the mean, mode, variance, etc., of the f-distribution.

The Student t-test and the variance ratio test may be applied mechanically. One should be more careful, however, when using the Chi-square test. To avoid statistical fallacies, the reader is advised to study the following examples, especially the Examples 11.6 and 11.7.

Example 11.6   Introduction to the Chi-square test

(A) An X-raying of insects resulted in twenty-four mutations from a sample of 2000 individuals (mutation rate: 1.2%). The normal mutation rate is 0.8% (sixteen out of 2000). Are the insects sensitive to radiation?

It is $k = 2$, $v = 1$, $\chi_{95}^2\% = 3.8$, $\chi_{99}^2\% = 6.6$.

	Mutations	No mutations	
X-rayed	24	1976	(f-distribution)
Not X-rayed	16	1984	(v-distribution)
Difference	8	8	
Yates-corrected	7.5	7.5	

Uncorrected: $(64/16)+(64/1984)=4$ $(4>\chi^2_{95}\%)$
Corrected: 3.5 $(3.5<\chi^2_{95}\%)$

The result is sensitive to the corrections of Yates. In other words, the result is sensitive to a very small change of the number of mutations. This indicates that the sample is too small.

(B) A group of 200 men suffering from a certain disease was treated with a medication; a second group of 200 men was not treated. Is the medication efficient? This was the outcome of the experiments:

	Cured	Not cured
Treated	168	32
Not treated	156	44

Strictly speaking, we may consider the results of the treated group as the f-distribution or as the v-distribution; from a scientific point of view, no group really is distinguished. In both situations the degree of freedom is equal to unity and the corrections of Yates should be applied.

f-distribution: treated cases: $\chi^2_s=3.9$
f-distribution: untreated cases: $\chi^2_s=4.9$ $\chi^2_{95}\%<\chi^2_s<\chi^2_{99}\%$

The first approach results in a value of χ^2_s that is equal approximately to $\chi^2_{95}\%$. Hence no decision should be made. According to the second approach, however, the medication may be efficient. Obviously, so far the outcome of the analysis is inconsistent and inconclusive. Let us therefore compare both groups simultaneously with the average outcome of the experiments (in the average, 162 men have been cured, 38 men have been left uncured):

	Contingency table of the f-distribution		Contingency table of the v-distribution		Table of difference	
Treated	168	32	162	38	6	6
Not treated	156	44	162	38	6	6
	(actual results)		(averages)			

$\chi^2_s=(30.25/162)+(30.25/38)+(30.25/162)+(30.25/38)=2.0$
$\chi^2_s=2.0<3.8=\chi^2_{99}\%$ $(v=1)$
$(30.25=5.5\times5.5=$ square of Yates-corrected differences$)$

The deviation is not significant and does not support the view that the medication really is effective. Hence the analysis indicates that the experiment should be repeated with a larger group of men. All in all, at the present stage of the investigation the medication does not seem to be effective.

(C) A pollen analysis resulted in the following figures (the example is simplified to avoid complications):

Sample	Sample size	Number of beechtree pollens
A	120	6 (5%)
B	80	8 (10%)
Sum	200	14 (7%)

Do the samples differ significantly? They cannot be compared directly because the sample sizes are different. We therefore test the hypothesis that no sample differs significantly from the average distribution which contains 7% beechtree pollens. The contingency tables are:

f-distribution

Sample	Beechtree	Not beech
A	6	114
B	8	72

v-distribution

Sample	Beechtree	Not beech
A	8.4	111.6
B	5.6	74.4

$$v = (2-1)(2-1) = 1$$
$$\chi_s^2 = 1.8 < 3.8 = \chi_{95}^2 \%$$

Hence the hypotheses that there is no significant difference cannot be refuted. For applications of statistics to pollen analysis see for instance Davis (1965) and Mosimann (1965).

Example 11.7   The effect of sample size and the number of classes on χ^2. Testing the significance of peaks

In the last examples, the Chi-square test was not as successful as expected because the number of individuals of the f-distribution, that is the sample size, N, was too small. We can expect better results if N becomes larger because doubling N results in doubling χ^2 if the relative frequencies are not effected by changing the sample size.

Example:

f:	5	6	10	12
v:	4	4	8	8

$$\chi_s^2 = 1.25 \qquad\qquad \chi_s^2 = 2.5$$

(The values of v are doubled and the squares of the differences f-v are quadrupled.) This indicates that the Chi-square test can be applied successfully if the sample size is sufficiently large.

As an example we consider a distribution of directions which exhibit two rather weak maxima. They are in the classes "20–40°" and "120–140°" respectively (zero degrees=north). We suspect that the maxima display a local phenomenon which is not typical for the region. We therefore test the hypothesis that every direction has the same frequency.

$N=90$ (number of measurements) $v=9-1=8$

Class	0°–20°	20–40	40–60	60–80	80–100	100–120	120–140	140–160	160–180
f-distribution	8	14	11	9	8	10	14	9	7
v-distribution	10	10	10	10	10	10	10	10	10
Differences	2	4	1	1	2	0	4	1	3

$N=$ 90: $\chi_s^2 = 5.2 = \chi_{26}^2\%$

$N=180$ (all f-values doubled): $\chi_s^2 = 10.4 = \chi_{75}^2\%$

$N=270$ (all f-values tripled): $\chi_s^2 = 15.6 = \chi_{95}^2\%$

Hence, the difference becomes significant for a sample size of approximately $N=300$. Now the number of classes is reduced to two. The first class contains all previous classes where f is larger than 10, the second class contains the rest. There results a *new* distribution that differs from equipartition:

Class	20–60; 120–140	0–20; 60–120; 140–180	
f	39	51	$N=90$
v	30	60	
d	9	9	

$$\chi_s^2 = 3.6 = \chi_{94}^2\%$$

Example 11.8   Comparing a distribution with a binomial distribution

The following refers to Example 9.1. The f-distribution is constituted by observations; the v-distribution is the binomial distribution with the same mean:

	$E_1(x=0)$	$E_2(x=1)$	$E_3(x=2)$	$E_4(x=3$ or 4 or 5)
f	12	22	8	18
v	6.44	18.12	20.38	15.05
d	5.56	3.88	12.38	2.95

$$\chi_s^2 = 13.7 \qquad k=4 \text{ classes}$$

How many parameters, h, of the v-distribution had to be estimated to calculate its frequencies? As every binomial distribution, the v-distribution is determined by two parameters: sample size and mean. The sample size is fixed, the mean had to be computed from the f-distribution. Hence $h=1$ and $v=k-1-h=4-1-1=2$. Therefore $\chi^2_{99}\%=9.2$ and $\chi^2_s>\chi^2_{99}\%$: the empirical distribution differs significantly from the corresponding binomial distribution. (There may exist, however, another binomial distribution with a different mean which does not differ significantly from the empirical distribution!)

Example 11.9 Comparing a profile with an idealized cycle (for details see Chapter 8, Example 8.1)

Actual profile (f-distribution):	4	45	10	32	9	33	23	110
Idealized cycle (v-distribution):	0	49	0	42	0	42	0	133

$k=8$ $\qquad$ $v=7$ (no parameter has to be estimated)

The details of calculation are left to the reader.

Example 11.10 The effect of rock composition on crystal form

Niggli (1948, p. 118) investigated three samples of garnet minerals with the faces (110) or (211) in the presence of quartz. All minerals were from druses, that is mineral-filled cavities or vugs, of a garnet-diopside rock from Maigels (Gotthard, Switzerland).

No.	Quartz	Number of garnets	Garnets with face (110) only
1	Insignificant	450	12 (2.7%)
2	Insignificant	526	13 (2.5%)
3	Rather abundant	381	61 (16 %)
	Sum	1357	86 (6.3%)

Is the relative abundance of the rhombic dodecahedron (110) of the last sample significant?

No.	f-distribution (110), (211)	(110) only	No.	v-distribution (110), (211)	(110) only
1	438	12	1	422	28 (6.3% of 450)
2	513	13	2	493	33 (6.3% of 526)
3	320	61	3	357	24 (6.3% of 381)

$\chi^2_s=83.6$
$v=(2-1)(3-1)=2$ $\chi^2_{99}\%=9.2$
$\chi^2_s>\chi^2_{99}\%$

Hence, the relative abundance of the rhombic dodecahedron is highly significant.

Example 11.11 Hunting the "forest elephant" in the Paleolithic (comparison of death-rate tables)

A limestone tuff from the younger Pleistocene at Taubach in the province of Thuringia (located between Bavaria and Saxonia) contains bones (most of them cracked) and single molars of *Elephas antiquus* which is supposed to have lived mainly in the woods. According to Soergel (1922, p. 78–128), these fossil remains are proof of successful hunting by man during the last interglacial. The proof rests on a comparison of the age distribution of Taubach's "forest elephants" with the age distribution of the elephants from the steppe (prairie) near the town of Suessenborn. We may assume that the elephants from the steppe died a natural death. (An expert can estimate rather easily the age of an elephant using the results of well-known studies of the recent Indian elephant. Molars are wornout and shortened in the course of time due to a grinding down of the teeth.)

Age	0–6	6–20	20–50	50–...	Sample size
Taubach	25.5%	28.8%	28.8%	16.7%	64
Suessenborn	0%	8.6%	13.4%	78.0%	200+?

The sample size of the group from Suessenborn is not well known but is larger than 200 animals. We therefore must reduce this group to 64 animals, that is to the size of the group from Taubach. Otherwise it would be difficult to apply the Chi-square test.

Age	0–20	20–50	50–...	
Taubach	35	18	11	*f*-distribution
Suessenborn	5	9	50	*v*-distribution

$$\chi_s^2 = 219 > 9.2 = \chi_{99}^2\% \quad (v = 3 - 1 = 2)$$

Hence, the difference of the two distributions is highly significant, even when reducing the sample size of the *v*-distribution to 64 animals (Adam, 1966).

Example 11.12 Comparing an empirical distribution with the corresponding normal distribution

The distribution shown in Chapter 9, Figure 22, subfigure 2 is for a sample size of 125 *Ceratites* with a mean of 75 and a variance of 44.

Hence we may compare it with a normal distribution of the same mean and variance. We have to use the approach indicated at Example 9.8 to obtain for each class the percentage values of the normal distribution. Finally, all values must be multiplied by $125/100 = 1.25$ to refer them to the sample size 125 (percentages refer to a sample size of 100). The empirical f-distribution extends from $x = 55$ to $x = 90$, whereas the normal distribution covers the entire interval from minus to plus infinity. Hence the class "55 ... 65" of the empirical f-distribution corresponds to the class "$-\infty$ ... 65" of the normal distribution, and the class "58 ... 90" corresponds to the class "85 ... ∞

Class	55–65	65–70	70–75	75–80	80–85	85–90
Empirical f-distribution	8	27	30	27	27	6
Normal distribution (v)	8.25	20	34.25	34.25	20	8.25

$$\chi_s^2 = 7.6$$

To calculate the normal distribution, we had to estimate two parameters, mean and variance. Hence $h = 2$ and $v = 6 - 1 - 2 = 3$ and $\chi_{95}^2\% = 7.8$. This value is close to χ_s^2 the sample seems to be too small to justify any decision.

Example 11.13   Investigating rock varieties and local fluctuations

Most geological bodies are rather inhomogeneous: its properties are different at different locations. Thus, when comparing rock specimen of a rather similar petrographic character, it may be difficult to decide whether they belong to the same type or whether they represent different varieties. With other words: where is the boundary between local fluctuations and varieties?

There are several possibilities in answer to this question. In every situation, one should first get N samples from the same rock variety. Then for each sample a mean is calculated. The set of N means constitutes a new sample whose mean and confidence interval can be calculated by applying the Student-t test. We then may postulate that every mean from an additional sample which falls outside of the confidence interval belongs to a different variety.

We also may apply the Chi-square test. Each sample constitutes a f-distribution (a contingency table). There is a total of N f-distributions (N contingency tables). We calculate the average distribution which is considered as the v-distribution. Then we compare each f-distribution (each contingency table) to the v-distribution (v-contingency table) thereby obtaining N values of χ_s^2 with a largest value χ_{max}^2. No χ_s^2-value of an additional sample representing the same variety should be larger than χ^2_{max}. Keep in mind that N should be fairly large and that all samples must have the same sample size or must be reduced to the smallest sample size of all samples.

11.4   Testing Hypotheses About Sequences of States

The first three examples presuppose some knowledge of Chapter 7.

Example 11.14   A simple test for inhomogeneity

We want to check the inhomogeneity of some quality arranged on an area. The area will be covered by a square grid. For each grid cell, the value of the property is determined. This results in a distribution with quartile Q_1. Now a cell is marked with a minus sign if the cell value is larger than Q_1. Otherwise the cell is marked with a plus sign resulting in a sequence of two states ($-$ or $+$).

Definition of symbols:

n_+, n_-	number of signs in a row of cells
$n = n_+ + n_-$	
u	actual number of runs in a row of cells
$\bar{u}_{id} = 1 + 2(n_+ n_-/n)$	expected number of runs of a random sequence of n_+ plus signs and n_- minus signs (cf. Chapter 7)

	n_+	n_-	n	u	$\bar{u}_{id}$
$-+---+--$	2	6	8	5	4
$------++-$	2	7	9	3	4.1
$---+---++---$	3	9	12	5	5.5
$+-----+++--$	4	7	11	4	6.1
Sum	11	29	40	17	19.7
Mean				4.25	4.9

There is inhomogeneity with respect to the first quartile Q_1 if the means 4.25 and 4.9 differ significantly. The actual mean 4.25 refers to a sample of size 4 ($N=4$ rows) whereas the mean 4.9 refers to a random theoretical distribution. Thus, one should apply case 1 of the Student-t test.

Example 11.15   Testing the significance of trend deviations

Let us consider the following time series:

Time	1	2	3	4	5	6
Actual values	0.1	1.4	3.8	8	16.5	24.4
Trend	0	1	4	9	16	25
Difference	+0.1	+0.4	−0.2	−1	+0.5	−0.6
Sign of difference	+	+	−	−	+	−

Sum of differences: −0.8   Mean difference: −0.8/6 = −0.13

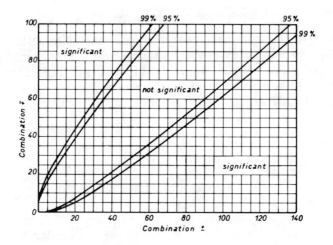

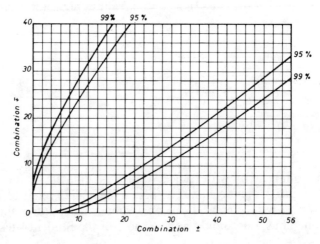

FIGURE 28. Diagram of Bertram (slightly simplified). Examples: If there are forty combinations "plus-minus" and twenty combinations "minus-plus" difference of corresponding sequences probably is significant. If numbers of combinations are 100 and twenty respectively, the difference is significant.

The sum of differences is rather small. The deviations from trend may be considered as random if the sequence of signs $++--+-$ does not differ significantly from the corresponding random sequence of plus and minus signs. For checking, the Student-t test (case 1) may be used. Thereby $\bar{x}$ is the mean run length of the actual sequence, and $\bar{X}$ is the expected theoretical run length of the random distribution which has the same number of plus signs and of minus signs as the actual sequence. N is the number of actual runs ($N=4$); s is the corresponding standard deviation.

Example 11.16   The Wald-Bertram test (Bertram, 1956, 1960)

The Wald-Bertram test seemingly is the simplest for comparing two sequences. The sequences

$$++-+--+$$
$$++-+--+$$

do not differ at all. The sequences

$$+--+--+$$
$$++-----+$$
(No. 1234567)

display two deviations: one combination "minus-plus" (No. 2), and one combination "plus-minus" (No. 4). There is a slight difference in arrangement, but the trend and the total number of plus signs (3) and of minus signs (4) are the same for both sequences. On the other hand, the sequences

$$++-+--+$$
$$--+-++-$$

show the maximum number of deviations, namely four combinations "plus-minus" plus three combinations "minus-plus."

The Wald-Bertram test discards all combinations "plus-plus" and "minus-minus" because they do not contribute to the number of deviations between two sequences of plus and minus signs. The test compares only the number of combinations "plus-minus" to the number of combinations "minus-plus". If they are equal, both sequences have the same trend and the same number of signs of the same types. If they are unequal, the sequences may differ significantly.

Example:

```
+ – – – + – + – – + + + + – + – –
+ – – + + – + + – + + – – + – + +
```

Total number of signs per sequence:	17 (discarded)
Number of combinations "plus-minus":	3
Number of combinations "minus-plus":	5

Thus the test does not compare the total number of signs with the total number of combinations of unequal signs! The number of terms of two sequences may be very large, and the number of deviations may be small. Nevertheless, the deviations compared among themselves may display a significant difference.

CHAPTER 12

Ranking Methods

Ranking methods permit the comparison of distributions whose absolute frequencies are not known quantitatively or do not exist. In this situation, classes are ordered according to their relative abundancies. Rank 1 is allocated to the class with the highest frequency, rank 2 refers to the class with the second largest frequency and so on. This approach is useful especially in the following situations:

1. Frequencies are described rather vaguely as "rare," "frequent," "abundant," etc.
2. It is too time-consuming to determine numerical frequencies by a number count.
3. Frequencies are relative and cannot be converted to absolute values, or the conversion to absolute values is useless.
   Example: The frequencies of a grain-size distribution are stated as weight percentages. The conversion to actual weights is not helpful because weights as well as volumes, areas, and lengths do not represent the number of items of a class.

To compare two distributions by ranking, the method of Spearman may be used. The details are explained in the following examples.

Example 12.1   Ranking of qualitative distributions: The degree of cementation in a diagenetic facies

The relative volumes of $k=6$ cementation minerals are estimated in two thin sections from different locations. The minerals are quartz, albite, anhydrite, calcite, dolomite, and barite. In the first thin section, the little differing frequencies of its rarest cementation minerals, Ca, Do, and Ba, get the same rank 5 which is the mean value of the ranks 4, 5, and 6.

Cement	Qu	Al	An	Ca	Do	Ba
Thin section 1	1	2	3	5	5	5
Thin section 2	4	3	5	1	6	2
Differences	−3	−1	−2	4	−1	3
Squares	9	1	4	16	1	9

Sum of squares: $40=S$

Now we compute the *Spearman's Rank Correlation Coefficient, R*, which is defined as

$$R = 1 - 6S/(k^3 - k)$$

Thereby k is the number of classes or ordered items and S the sum of squared rank differences. We obtain

$$R = 1 - (6 \times 40)/(216 - 6) = -0.14.$$

R is equal to unity ($R = 1$) if the ranked distributions coincide, and $R = -1$ if the rank sequences run opposite.

Example:

First rank distribution:	1 2 3 4 5 6 7	
Second rank distribution:	1 2 3 4 5 6 7	$R = 1$
First rank distribution:	1 2 3 4 5 6 7	
Second rank distribution:	7 6 5 4 3 2 1	$R = -1$

A rank coefficient that is slightly negative as in the example ($R = -0.14$) indicates that the compared distributions differ distinctly.

Example 12–2   Ranking of relative frequency distributions: Chemical composition of basalts

We now consider an adaptation of ranking to relative frequencies which sum-up to 100%. Let k be the number of classes of each distribution. Then the *ranking sum* is defined as $K = 1 + 2 + 3 + \ldots + k$ (*Example: $k = 5$: $K = 1 + 2 + 3 + 4 + 5$*). When multiplying each relative frequency with $K/100$, there result new distributions with total cumulative frequencies K rather than 100. Finally, we calculate the Rank coefficient R for the new distributions.

Chemical composition of plateau basalts, average values (%) Barth (1939, p.71)

	SiO_2	TiO_2	Al_2O_3	Fe_2O_3	FeO	MnO	MgO	CaO	Na_2O	K_2O	H_2O	P_2O_5
Deccan	50.54	1.91	13.56	3.19	9.91	0.16	5.45	9.44	2.60	0.72	2.13	0.39
Earth's crust	47.14	2.44	14.91	4.11	8.22	0.25	6.91	10.01	2.71	0.84	2.13	0.33

Sum of relative frequencies: 100; $k = 12$ classes

$$K = 1 + 2 + 3 + 4 + \ldots + 12 = 78$$

Each frequency is multipled with 0.78:

Deccan	39.4	1.5	10.6	2.5	7.7	0.1	4.3	7.4	2.0	0.6	1.7	0.3
Earth's Crust	36.8	1.9	11.6	3.2	6.4	0.2	5.4	7.8	2.1	0.7	1.7	0.3
Differences	2.6	−0.4	−1	−0.7	1.3	−0.1	−1.1	−0.4	−0.1	−0.1	0	0
Squares	6.76	0.16	1	0.49	1.69	0.01	1.21	0.16	0.16	0.01	0.01	0 0

Sum of squares $S = 11.5$
$R = 1 - (6 \times 11.5)/(1728 - 12) = 0.96$

Thus, the deviation of the chemical composition of the Deccan basalts from the worldwide average composition is represented by a single number which indicates that the difference is small. When using the complete tabulation of Barth, the following values are obtained. They indicate a high degree of homogeneity of basalts:

Region	R
Deccan (India)	0.96
Oregon	0.93
Arctic	0.99
Scotland	0.99
New Jersey	0.97
Mean	0.97

Example 12.3   Comparing the chemical composition of two types of metorites

	Ni-Fe	Olivine	Pyroxenes	Feldspars		Ranking		
Achondrite	1.6	12.8	62.3	20.8%	4	3	1	2
Chondrite	10.6	42.3	29.9	11.8%	3.5	1	2	3.5

$R = 0.25$.

To compare simultaneously more than two distributions by the method of ranking, the *coefficient of concordance*, W, may be used. W is equal to unity ($W = 1$) if all distributions coincide. If, however, all ranks are distributed randomly, W is equal to zero.

Example 12.4   The comparison of m rank sequences

Let us consider three distributions ($m = 3$). For instance, they may represent three different geological profiles or certain properties of these profiles. Each distribution consists of $k=9$ classes. The relative frequency of each class is indicated by ranking. First, the numbers

$$M=m(k+1)/2 \text{ and } A=m^2(k^3-k)/12$$

are calculated: $M=15$, $A=540$.

Class	I	II	III	IV	V	VI	VII	VIII	IX
1st distribution	9	6	3	4	1	2	5	7	8
2nd distribution	1	3	6	4	2	5	7	8	9
3rd distribution	6	2	1	4	3	7	9	8	5
Sums	16	11	10	12	6	14	21	23	22
M	15	15	15	15	15	15	15	15	15
Differences	1	−4	−5	−3	−9	−1	6	8	7
Squares	1	16	25	9	81	1	36	64	49
Sum of squares:		$S=282$							

Then

$$W=S/A: \qquad W=0.52$$

Testing the significance of R. The number of classes, k, must be equal to ten or larger. Calculate $B=(k-2)/(1-R^2)$ and $t_o=R$ times the square root of B. Compare t_o with the value of t from Table III of the Appendix for $v=k-2$ degrees of freedom. R is significant at the 1% (5%) level, if t_o is larger than the table value $t_{99\%}$ ($t_{95\%}$).

Testing the significance of W. Compute $W^*=(S-1)/(A+2)$ and $F_o=(m-1)W^*/(1-W^*)$. Apply the Snedecor F-test by comparing F_o to the F-*value* of Table IV of the Appendix for the degrees of freedom

$$v\text{-column} \quad = (k-1)-(2/m)$$
$$v\text{-row} \qquad = (m-1) \text{ times } v\text{-column.}$$

W is significant if F_o is larger than the table value. Is the value of W of Exercise 12.4 significant at the 5%-level? ($W^*=0.518$; $F_o=2.15$; v-column $=7.33$; v-row $=14.67$; $F=2.6$ (estimated table value). Therefore, the significance of W is in doubt.)

CHAPTER 13

Correlation and Regression

Up to now we have considered mainly the variation of a single quality or quantity (*univariate distribution*). In many situations, however, there are several properties which differ simultaneously such as the lengths and the breadths of pebbles (*multivariate distribution*).

Let us first consider *bivariable distributions* with variable properties denoted by the letters x and y. Such distributions are represented easily by a *contingency table*. Each box of the table displays the absolute frequency of the simultaneous occurrence of a certain x-value or x-class and a certain y-value or y-class. For instance, in the following table five pebbles have the length $x=3$ and the breadth $y=2$, and eight pebbles have the length $x=5$ and the breadth $y=3$.

	6	10	24	12	1	f_x / f_y
5	0	1	3	2	0	6
4	0	4	6	4	0	14
3	1	2	8	5	1	17
2	5	3	7	1	0	16
y / x	3	4	5	6	7	

The top row of the scheme contains the frequency distribution of the x-values regardless of the y-values. For instance, $5+1+0+0=6$ pebbles have the length $x=3$, and $3+2+4+1=10$ have the length $x=4$. Thus, the top row is a univariate distribution of x termed the *marginal distribution* f_x of the corresponding bivariate distribution. There is a second marginal distribution f_y where the x-values are discarded. This is a univariate distribution of y-values represented by the last column. For instance, $0+1+3+2+0=6$ pebbles have the breadth $y=5$.

To obtain the corresponding relative bivariate and marginal frequency distributions, all absolute frequencies must be divided by the sample size N, that is by the total number of pebbles;

$$(N=6+10+24+12+1=6+14+17+16=53).$$

A bivariate distribution may indicate a *trend*, that is a tendency that a change of the property x corresponds to a change of the property y. The *goodness of a trend* can be estimated from the *product moment correlation coefficient*, r. The calculation of r is rather cumbersome, however, but an explanation on the determination of r will be given here avoiding any advanced mathematical notation.

The step-by-step calculation of the correlation coefficient, r

Let the bivariate distribution be

	5	10	5	f_x / f_y
4	4	8	4	16
2	1	2	1	4
y / x	1	2	3	

1st step: Calculate the mean values $\bar{x}$ and $\bar{y}$ of the marginal distributions f_x and f_y respectively:

x:	1	2	3		y:	2	4
f_x:	5	10	5		f_y:	4	16

$$\bar{x}=2 \qquad\qquad \bar{y}=3.6$$

2nd step: Replace x by x-$\bar{x}$, and replace y by y-$\bar{y}$:

0.4	4	8	4
−1.6	1	2	1
$y-\bar{y}$ / $x-\bar{x}$	−1	0	1

3rd step: Multiply the frequency of each cell with the corresponding values of $x-\bar{x}$ and $y-\bar{y}$:

−1.6	0	+1.6
1.6	0	−1.6

4th step: Compute the sum of the new cell values. (In this example, this sum is zero.)

5th step: Divide the sum by the sample size N, that is by the sum of all frequencies (in this example, $N=20$). The result is termed the *covariance*, μ_{11}. (In this example, the covariance is equal to $0/20=0$.)

6th step: Calculate the standard deviations σ_x and σ_y of the marginal distributions f_x and f_y respectively.

7th step: Compute $r=\mu_{11}/(\sigma_x\sigma_y)$.

It may be shown that the correlation coefficient, r, cannot exceed $+1$ or be less than -1. The values of $+1$ and -1 denote a perfect linear relationship between x and y. In this situation, when plotting all x,y-pairs of the distribution on a rectangular x,y-coordinate system, all plotted points are located strictly on an ascending ($r=+1$) or descending ($r=-1$) straightline (cf. Figure 29-1). If r is less than $+1$ but positive, the linear relationship is somewhat blurred. There is, however, a trend that increasing x is associated with increasing y. This is termed *positive correlation* (Figure 29–3). If r is negative there is a *negative correlation*

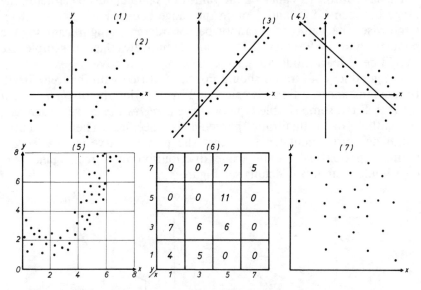

FIGURE 29. Plot of correlations. There are N objects with two properties, x and y (bivariable distribution with sample size N). Each object is associated with one value of x and one value of y. x and y define one point on rectangular coordinate system. (1) All points are on straightline: linear relationship between x and y. (2) All points are on curved line: nonlinear relationship between x and y. (3) Positive linear correlation. (4) Negative linear correlation. (5) Nonlinear trend. (6) Contingency table associated with graph (5). (7) Random distribution of points: no trend, no correlation.

(Figure 29–4). The value $r=0$ indicates that there is no trend at all. In this situation the plotted points are distributed randomly (Figure 29–7). The correlation coefficient also is zero if all plotted points are located on a parallel to the x-axis or the y-axis. In these situations, too, there is no trend. This may be confirmed by sketching corresponding plots.

The discussion shows that a linear trend is pronounced if r is near to $+1$ or to -1. On the other hand, a linear trend is more blurred as r approaches zero. If a trend is *nonlinear*, the correlation coefficient will always be different from $+1$ or -1. Strictly speaking, r refers to linear rather than to more general correlations.

A trend may indicate a *causal relationship* but not necessarily so. Even in the situation of $r=1$ or $r=-1$, the relationship may not be the effect of any physical, chemical, biological or other known causation (then it is a "nonsense correlation").

In general, the correlation coefficient of a sample differs from the value of r of the universe. The sample may indicate goodness of trend which is not justified for the population. To check the reliability of r for *linear* correlation, Table VI of the Appendix may be used.

Example: Sample size: $N=5$ (five objects with properties x and y).

Then according to Table VI, the value of r-sample must be either in the range between 0.878 and 1 or in the range between -1 and -0.878. Otherwise, the value of r cannot be considered as significant for the population at the 95% level. The example indicates that the sample size should be at least moderately large to be representative.

The goodness of a trend should not be confused with the trend itself. In most instances, a trend is calculated by applying the *method of least squares*. First, some simple type of curve ("*regression curve*") such as a straightline or a quadratic parabola is specified. Then the curve parameters are calculated such that the plotted curve is close to the points representing the bivariate distribution in a rectangular x,y-coordinate system (for details see Figure 30).

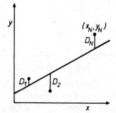 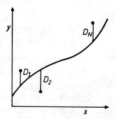

FIGURE 30. Method of least squares. D_N is distance of nth point to regression curve (recorded in direction of ordinate y). Sum of squares $D_1^2 + D_2^2 + D_3^2 + \ldots + D_N^2$ is minimized. Value of the sum depends on specified curve type.

Linear regression lines

Equation of the regression line: $y=\bar{y}+(\mu_{11}/\sigma_x^2)(x-\bar{x})$

$\bar{x}, \bar{y}$: the means of the marginal distributions
σ_x^2: variance of the marginal distribution of x-values

If the covariance vanishes ($\mu_{11}=0$, i.e., $r=0$), the regression equation degenerates to $y=\bar{y}$. This denotes that y is equal to $\bar{y}$ for every value of x. In other words, y is not correlated to x at all.

Numerical example: $\bar{x}=4$, $\bar{y}=2$, $\sigma_x^2=3$, $\mu_{11}=9$: $y=2+3(x-4)=3x-10$.

Remark: The method of least squares distinguishes the property y as is indicated at Figure 30. Actually, it is completely at the discretion and the judgement of the user which property is labeled as y and which is denoted by x.

The regression line represents a best estimate of the trend but does not indicate the possible range of trend values. If $y=\bar{y}+(\mu_{11}/\sigma_x^2)(x-\bar{x})$ is the regression line of a sample, then the regression line of the universe is in the range

$$y=\bar{y}+S+(\mu_{11}/\sigma_x^2)(x-\bar{x})$$

and

$$y=\bar{y}-S+(\mu_{11}/\sigma_x^2)(x-\bar{x})$$

where S is a constant that depends on several factors. If the sample size N is not too small (say N is at least 30), and if both marginal distributions are approximately normal distributions, then S may be estimated easily. Calculate the standard deviation of the marginal distribution of y-values and multiply it by the square root of $1-r^2$. Then S is equal to this product times 2 at the 95% level and equal to this product times 3 at the 99.9% level.

Quadratic regression

The trend of the distribution is described by a quadratic parabola $y=a_o+a_1x+a_2x^2$. It has either a relative maximum or a relative minimum at $x=-a_1/2a_2$. Let the sample consist of the N pairs of numbers x_1,y_1 x_2,y_2 $x_3,y_3 \ldots x_N,y_N$. The constants a_0, a_1, a_2 of the parabola are the solution of the system

$$[y]=a_0 N+a_1 [x]+a_2 [xx]$$
$$[xy]=a_0 [x]+a_1 [xx]+a_2 [xxx]$$
$$[xxy]=a_0[xx]+a_1 [xxx]+a_2 [xxxx]$$

where

$$[x]=x_1+x_2+\ldots+x_N \qquad [y]=y_1+y_2+\ldots+y_N$$
$$[xx]=x_1{}^2+x_2{}^2+\ldots+x_N^2 \qquad [xy]=x_1y_1+x_2y_2+\ldots x_Ny_N$$
$$[xxx]=x_1{}^3+x_2{}^3+\ldots+x_N^3 \qquad [xxy]=x_1{}^2y_1+x_2{}^2y_2+\ldots+x_N^2y_N$$
$$[xxxx]=x_1{}^4+x_2{}^4+\ldots+x_N^4$$

Linear regression for three variable properties

If a distribution depends on three random variables, x, y, and z, the trend may be described by the *regression plane* $z=a_0+a_1x+a_2y$. The constants a_0, a_1, and a_2 must be determined by solving the equations

$$[z] \ =a_0N+a_1[x]+a_2\,[y]$$
$$[xz]=a_0\,[x]+a_1\,[xx]+a_2\,[xy]$$
$$[yz]=a_0\,[y]+a_1\,[xy]+a_2\,[yy]$$

To check the goodness of the trend, one calculates first the correlation coefficients r_{xy}, r_{yz}, and r_{zx}. The use of these symbols is best explained by a simple numerical example. Let the distribution of three variables x, y, and z be

	z=1:						z=2:		
3	1	2	8			3	3	2	0
2	5	3	7			2	4	4	0
y/x	3	4	5			y/x	3	4	5

Then the bivariate distribution of x and y *irrespective of z* is

3	4	4	8	because	1+3=4	2+2=4	8+0=8
2	9	7	7		5+4=9	3+4=7	7+0=7
y			
x	3	4	5

r_{xy} is the correlation coefficient of this contingency table. In other words, r_{xy} is the correlation coefficient of the marginal distribution which results from the original distribution by eliminating z.

The original distribution may be rewritten in the form

	y=2:						y=3:		
2	4	4	0			2	3	2	0
1	5	3	7			1	1	2	8
z/x	3	4	5			z/x	3	4	5

Hence, the bivariate distribution of x and z *irrespective of y* is

| 3 | 7 | 6 | 0 |
| 2 | 6 | 5 | 15 |

$$\frac{z}{\diagup x} \quad \begin{array}{ccc} 3 & 4 & 5 \end{array}$$

because $4+3=7$, etc.

Then r_{zx} is the correlation coefficient of this contingency table. Finally, $r_{xy.z}$, the *partial correlation of x and y*, is calculated. It shows the goodness of trend of x with respect to y after eliminating the effect of the variation of z. We calculate

$$A = r_{xy} - r_{yz}r_{zx}$$
$$B = \text{square root of } 1 - r_{yz}^2$$
$$C = \text{square root of } 1 - r_{zx}^2$$
$$D = B \text{ times } C$$
$$r_{xy.z} = A/D$$

Table VI may be used to check the significance of $r_{xy.z}$. In this situation the letter N of the table signifies the actual sample size minus 1.

All calculations that refer to the method of least squares, regression, and correlation should be performed with a large number of significant figures because even small rounding errors may have a large effect on the final results.

Example 13.1   Petrophysical correlations

Figure 31 shows typical linear correlations of the logarithm of permeability to air vs. other important rock parameters (Tunn, 1966). There also may occur quadratic regression curves with a maximum.

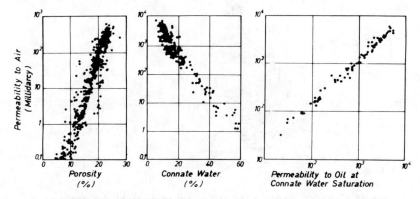

FIGURE 31. Log of permeability to air vs. porosity, connate water saturation, and log of permeability to oil at connate water saturation respectively.

Example 13.2   Determining the grid constants of cubic minerals by using the method of Taylor, Sinclair, Nelson, and Riley

The grid constant, a, is determined for different values of the planes (hkl) from the corresponding glancing angles, θ, using the powder method of Debye. There is a linear correlation between the grid constant, a, and the expression $[(1/\sin\theta + (1/\theta)]\cos^2\theta$. Extrapolation of the regression line furnishes an accurate value of a. (For details see Azároff and Buerger, 1958).

Example 13.3   Longshore currents

Usually, sea waves are not parallel to the shoreline. As a consequence, there results a zigzag current near the beach. Its observed velocities are correlated well with the values calculated from the theory of Putnam and others (1949). (For details see Scheidegger, 1961, p. 205 and fig. 111).

Example 13.4   The dependence of the number of species on climate

In general, the number of the species of a genus increases if the average temperature of January or July becomes larger (Figure 32). As a

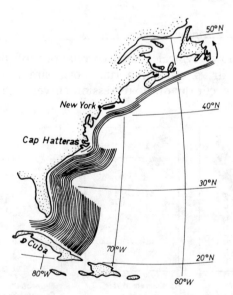

FIGURE 32. Eastern coast of USA and Canada: Geographical distribution of number of species of gastropods. Each line corresponds to ten species.

consequence, the geographic distribution of fossil species may infer the shifting of climate zones during the past (Fischer, 1960; Schwarzbach, 1963, p. 33).

Example 13.5 The discontinuous change of a specific angle of the Foraminifera Vaginulina procera *in the Barrême of Northwestern Germany*

When plotting the mode of a specific angle of *Vaginulina procera* vs. depth, Albers (1952) observed a jump of the modal values at geode layer (Figure 33-1, 33-2, 33-3). Albers thought that this corresponded to a hiatus or to a considerable slowing-down of sedimentation. This assumption was confirmed by Bettenstaedt (1962) who discovered the missing layer and the corresponding missing link in the evolution of the Foraminifera. Evolutionary jumps also are known within layers of continuous sedimentation. They may be the consequence of migration of races (Kaufmann, 1933, p. 50).

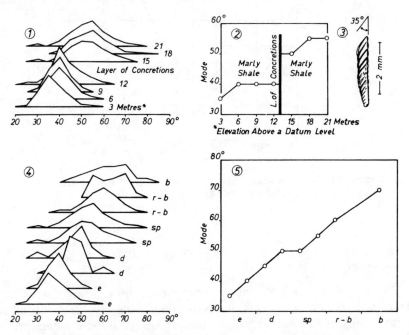

FIGURE 33. Correlation between specific angle of Foraminifera *Vagulina procera* and depth in Barrême profile of northwestern Germany. (1) Distribution curves for various depths from single profile with hiatus. (2) Corresponding diagram of mode vs. depth (Albers, 1952). (3) Definition of specific angle. (4) Complete system of distribution curves for various depths combined from several profiles (e=*elegans*, d=*denckmanni*, sp=*sparsicosta*, r-b=*rude-bidendatum* b=*bodei*). (5) Corresponding diagram of mode vs. depth (Bettenstaedt, 1962).

*Example 13.6   The significance of envelopes of widely
scattered points*

Not much information is conveyed where little if any correlation
exists between the two properties *x* and *y*, of a number of samples. In this
situation the points of a corresponding *x,y*-diagram are scattered widely.
In general, however, there is a curve which is an envelope of all points. It
may happen that this curve is representative of the considered
population. Figure 34 shows as an example the envelope for the medians
of sandstones plotted against the sorting coefficient. It also may be the
situation that the envelope of the plot of median vs. the clay content is
characteristic (personal communication from H. Füchtbauer).

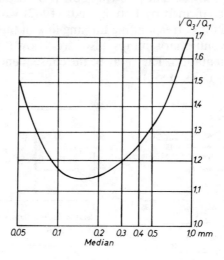

FIGURE 34. Lower envelope of sorting coefficient of natural sediments.
(Füchtbauer and Müller 1970; fig. 3–16).

*Example 13.7   The correlation between depth and
composition of crudes*

During geological time, as a consequence of tectonic forces and
deposition of new sediments, many oil reservoirs have been buried to
great depths. If this has happened, the pressure and the temperature of
the reservoir will have increased. This will have resulted in the
formation of additional lighter components which in general, could not
have escaped because of the compaction at greater depth. This implies
that the correlation between composition and maximal burial depth of a
reservoir should be better than the correlation between composition and
present depth (Phillip and others, 1969).

Example 13.8   The formation of paraffins in sediments: a
remark

In shallow-depth marine clays and silts, the normal paraffins with an
odd number of carbon atoms are more abundant than the paraffins with
an even number of C-atoms. This results in multimodal distributions of
n-paraffins. When rock samples are heated prior to the distillation of
hydrocarbons, the modes become less pronounced and the main
maximum shifts to lower C-values. This shows that the composition of
crudes is an indicator of the highest temperature to which the oil was
exposed in the past (for details see Welte, 1966).

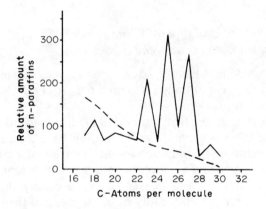

FIGURE 35. Distribution curves of n-paraffins distilled from oil shales.
Samples not heated: multimodal distribution. Sample heated to 490°C
dashed line.

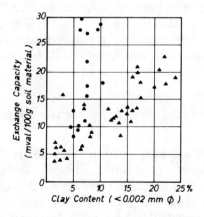

FIGURE 36. Exchange capacity vs. clay content for brown soils (triangles) and
loosely packed brown soils (circles) near Zwiesel (Brunnacker, 1965; and cf.
Example 13.9).

Example 13.9   The exchange capacity of soils: a remark

According to the regression analysis of Schachtschabel and Renger (1966), the exchange capacity for cations in soils depends essentially on the percentages of clay and humus and on the composition of these two substances. Figure 36 shows an example from brown soils near the northwestern Bavarian town of Zwiesel (Brunnacker, 1965). The exchange capacity is rather well correlated to the clay content of brown soils which do not seem to contain much organic material (cf. the triangles of Figure 36). Surprisingly, however, this does not hold for loosely packed brown soils (circles of Figure 36), which do not display any correlation between clay content and exchange capacity.

Example 13.10   The evolution of Kosmoceras in the middle Jurassic of England. A general trend-analysis test

The following method has the advantage that it can be applied to any type of regression curve although it is used here to check the significance of a straightline correlation.

In an excellent treatise, Brinkmann (1929) investigated evolutionary tendencies of *Zugokosmoceras*. Two conditions have to be satisfied for a consistent evolutionary trend such as the increase of diameter versus depth. First, the means of all sample values which belong to the same depth range should be close to the trend curve. Secondly, the mean of the lowest layer must differ significantly from the mean of the highest layer. This holds if the mean of the range 58–60 cm (sample size: nine individuals) is compared to the mean of the interval 76–78 cm (sample size: five individuals) by applying the Student *t*-test. However, in this example most confidence intervals overlap because for most layers the sample sizes are very small (one to five fossils) (Figure 37). For details see Marsal (1949). For a general mathematical treatment of growth problems see Simpson, Roe, and Lewontin (1960, chap. 15), and Miller and Kahn (1962, chap. 9).

Example 13.11   Comparing two linear correlations by using Snedecor's F-test

This example relates to a comparison of premolars P3 and P4 from the upper jaws of Pleistocene rhinoceros. All samples of teeth were derived from the same layer in a small region (data: K. D. Adam, pers. comm.). Figure 38 shows a plot of tooth breadth at the back versus tooth breadth at the front (sample sizes: P3: $N=6$ teeth, P4: $N=5$ teeth). Obviously, the correlation for the premolar P4 is superior to the correlation of the premolar P3.

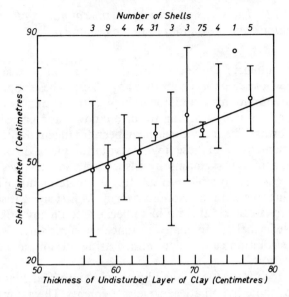

FIGURE 37. *Zugokosmoceras*, growth of diameter versus time. Data: see Brinkmann (1929, p. 60, table 21). Each mean which corresponds to thickness interval is indicated by circle. Vertical lines represent confidence intervals of means at 95% level.

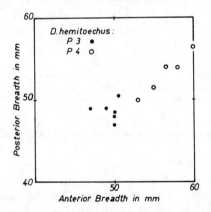

FIGURE 38. Correlation of breadths for premolars P3 and P4 of *Dicerorhinus hemitoechus*.

The distributions of tooth size are transformed to a distribution of a single variable by calculating for each premolar the ratio R=breadth at the front/breadth at the back. It turns out that the premolar P4 has a very small variance of R whereas the corresponding variance of the premolar P3 is seventeen times larger (the variances are 0.00006 and 0.00106 respectively). According to Snedecor's F-test, the variances differ significantly.

Example 13.12   The statistical reconstruction of fragmented fossil populations: A contribution to the taxonomy of conodonts

The paleontologist may be confronted with a blend of fossil conodont populations whose remains occur in many places. These populations are composed of individuals of unknown shape with an unknown number of skeleton elements which belong to an unknown number of unknown taxonomic forms. The individuals have been fragmented into a mixture of skeleton elements, and these fragments of individuals do not allow the reconstruction of the skeletons by inspection. So the question arises whether there is any possibility of splitting the material into classes, each class belonging to a unique taxonomic form. At first sight, it seems to be hopeless to try to tackle this problem because too many unknowns are involved. Nevertheless, in many instances a surprisingly simple and unambigious solution can be determined using the simple technique of linear correlation.

The basic idea is best explained by an oversimplified example which makes use of some hypothetical skeletal elements. The assumptions are refined at a later stage. Let there be two classes of individuals. The skeleton of the first class, I, is composed of one element a and two elements b. The skeleton of the second class, II, is composed of three elements c. Altogether, there are *three types* of skeleton elements, namely a, b, and c. Dismembered but otherwise fully preserved skeletons are present, at five different locations:

Sample	(1)	(2)	(3)	(4)	(5)	
Class I	2	3	6	4	5	skeletons
Class II	5	9	2	8	0	skeletons

Each sample is a mixture of elements. For instance, the first sample consists of 2 elements of type a, four elements of type b, and fifteen elements of type c. We begin the analysis by comparing each type of element with each other type. The comparison of a and b shows that at each location the number of b-elements is twice the number of a-elements. All points of the corresponding diagram are located on a straightline passing through the origin (Figure 39). On the other hand, the comparison of c with a and b shows that no simple linear relationship exists between the frequencies of a and c or b and c. This leads to the conclusion that the material consists of two types of skeletons, the first being composed of one element a and two elements b or a multiple of this combination, whereas the second skeleton is composed of three elements c or a multiple of three.

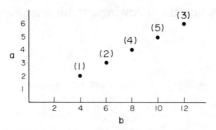

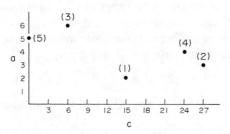

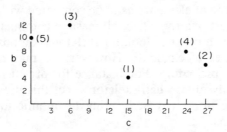

FIGURE 39. Comparison of frequencies of elements a, b, and c of samples (1) to (5) of last table.

Now let us consider a slightly more complex but by far more realistic example. There are six types of skeletal elements labeled a, b, c, d, e, and f, respectively. In the following scheme, the statement "Yes" (y) indicates that the ratio of the frequencies of the compared elements is the same for each sample; the statement "No" (n) indicates that the ratios differ for at least two samples.

	a	b	c	d	e	f
a		y	y	n	n	n
b			y	n	n	n
c				n	n	n
d					n	n
e						y

This scheme shows that the following combinations are feasible:

$$a,b, \qquad a,c \qquad b,c \qquad e,f$$

Now let

$$a{:}b = 1{:}2, \qquad a{:}c = 1{:}3, \qquad b{:}c = 2{:}3$$

Hence

$$a{:}b{:}c = 1{:}2{:}3$$

In this situation there are three types of skeletons characterized by the combinations of skeleton elements (a,b,c), (e,f), and (d) respectively. Thus, there are three different taxonomic forms, each characterized by a certain ratio of certain elements. If, however,

$$a{:}b = 1{:}2, \qquad a{:}c = 1{:}3, \qquad b{:}c = 1{:}1,$$

then a fixed ratio a:b:c does not exist and the combination (a,b,c) is not feasible.

In general, each sample may contain a large number of different types of elements. Because nothing is known about the affiliation of these elements, each type of element has to be compared with each other type looking for proportionality. Then, all feasible combinations have to be compared with each other to look for skeletons composed of more than two different types of elements. However, we may not get complete skeletons by this procedure. For instance for organisms with bilateral symmetry, we only obtain half-skeletons. Yet this drawback is not too serious if we are interested in locating taxonomic forms rather than complete skeletons.

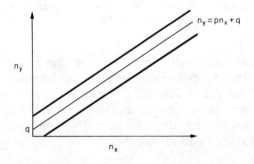

FIGURE 40. Regression line which intercepts abscissa at $n_y = q$ rather than at origin. Regression line is embedded in strip whose width may be estimated by using method explained in paragraph on linear regression lines.

The workability of the basic concept rests mainly on three assumptions:

1. Elements of different taxonomic forms are different in at least one recognizable respect. If this condition is violated, inconsistent proportions may arise and existing ratios may not be recognized. (*Example*: Different forms have in common the same type of element, a.)
2. At least two samples differ in the relative number of individuals of different taxonomic forms. For, if each individual of one form always is in company with the same number of individuals of another form, the analysis results in the construction of super-individuals possibly reflecting phenomena such as symbiosis.
3. All skeletons may be dismembered, but no elements are removed and no elements are added. This assumption usually is never met. There are two main types of errors leading to a violation of the last assumption: systematic and random errors.

Systematic errors rarely can be detected by a statistical analysis. Systematic errors may arise from the devouring of relatively soft elements by animals, from selective chemical dissolution of skeletal elements, and from other selective agents such as bottom currents on the sea floor. Errors also may be the result of the methods used in the laboratory such as fractionation of samples by sieves which tend to separate globular and spherical elements from needle-shaped and harpoon-like ones. Random errors that may result from various postmortem events tend to blur the strict linear relationship among elements of the same taxonomic form to a correlated linear trend. Without blurring plotting the numbers n_x of elements of the type x versus the numbers n_y of another type y, a straightline passing through the origin should result. In other words, a linear equation

$$n_y = pn_x + q \quad \text{where} \quad q = 0$$

should result. Actually, in general a regression line

$$n_y = pn_x + q \quad \text{where} \quad q \neq 0$$

will result, and the points of the plot will not be located strictly on a straightline but rather in a strip which contains the origin (Figure 40). If the sample sizes are large and if the correlation coefficient, r, is close to unity (say larger than 0.8), we may conclude that the ratio $n_y : n_x$ is nearly equal to p. This result, however, may not be unique. When plotting n_x along the abscissa and n_y along the ordinate, a regression equation $n_y = pn_x + q$ will result. When plotting n_y rather than n_x along the abscissa, a different equation $n_x = \bar{p}n_y + \bar{q}$ will result. The ratios $n_y : n_y = p$ and $n_y : n_x = 1/\bar{p}$ are only equal if $p = 1/\bar{p}$. In general, this might not be the situation. Yet uniqueness may result if more than two elements are considered simultaneously by combining for instance ratios a:b, a:c, b:c, a:d, and c:d to a common consistent ratio a:b:c:d.

Numerical examples:

(1) An "apparatus" with two elements:

$$a = -2.8 + 1.3b \qquad a{:}b = 4{:}3$$
$$b = 2.6 + 0.7a \qquad a{:}b = 3{:}2$$

(2) An apparatus with three elements:

$$a{:}b = 2{:}1 \text{ or } 9{:}5$$
$$a{:}c = 14{:}5 \text{ or } 3{:}1$$
$$\underline{b{:}c = 3{:}2 \text{ or } 5{:}3}$$
$$a{:}b{:}c = 6{:}3{:}2$$

For details see Marsal and Lindström (1972, p. 43–46).

CHAPTER 14

Simplified Linear and Nonlinear Regression Analysis

In this chapter two topics will be discussed: a simplified numerical approach to linear, quadratic, and cubic regression, and the reduction of important nonlinear problems to linear regression analysis.

The actual performance of a regression analysis involves a large number of numerical computations. Therefore, usually a computer with implemented statistical programs is employed. In most situations, however, the calculation routines can be simplified so that the computations can be performed in the field with a handy pocket calculator.

Let there be n pairs of data, x_1,y_1 x_2,y_2 $x_3,y_3 \ldots x_n,y_n$. We are looking for a least-square fit yielding one of the regression equations

$y=A_0+A_1x$ (linear regression)
$y=B_0+B_1x+B_2x^2$ (quadratic regression)
$y=C_0+C_1x+C_2x^2+C_3x^3$ (cubic regression)

In the following, the set of x-values must satisfy two conditions which are met in the majority of all instances. First, every x-value must occur exactly once. (i.e., there should be no pairs $x=2$, $y=6$ $x=2$, $y=3$). Secondly, the spacing between consecutive x-values ordered according to size must constant. $x_2-x_1=x_3$ x_2-x_4 $x_3-\ldots=x_n-x_{n-1}=h=$ const.

For instance, this condition is met by the sequence of x-values 1, 4, 7, 10 where $n=4$ and $h=3$. It is not met for the sequence 0, 1, 4, 10 with $n=4$ and the spacings 1, 3, 6. It further is not met by the sequence 1, 1, 1, 4, 7, 10 because the value $x=1$ occurs more than once.

The calculation routine

First step

If the number of pairs, n, is *odd*, calculate $m=\frac{1}{2}(n-1)$. This always is an even number. Then x is related to a new variable, u, by the following simple scheme:

x:	x_1	x_2	x_3	x_4	$\ldots x_n$
u:	$-m$	$-m+1$	$-m+2$	$-m+3$	$\ldots +m$

Numerical example: $n=5$, that is $m=2$

$$
\begin{array}{llllll}
x: & 1 & 4 & 7 & 10 & 13 & (h=3) \\
u: & -2 & -1 & 0 & 1 & 2 &
\end{array}
$$

If n is an *even number*, the following scheme is used:

$$
\begin{array}{llllll}
x: & x_1 & x_2 & x_3 & x_4 & \ldots\ x_n \\
u: & 1-n & 1-n+2 & 1-n+4 & 1-n+6 & \ldots\ n-1
\end{array}
$$

Numerical example: $n=4$

$$
\begin{array}{lllll}
x: & 1 & 3 & 5 & 7 & (h=2) \\
u: & -3 & -1 & 1 & 3 &
\end{array}
$$

Second step

Linear regression: calculate

$$
\begin{aligned}
[y] &= y_1+y_2+y_3+\ldots+y_n \\
[uy] &= u_1y_1+u_2y_2+u_3y_3+\ldots+u_ny_n
\end{aligned}
$$

Quadratic regression: calculate additionally

$$[uuy]=u_1^2y_1+u_2^2y_2+u_3^2y_3+\ldots+u_n^2y_n$$

Cubic regression: calculate additionally

$$[uuuy]=u_1^3y_1+u_2^3y_2+u_3^3y_3+\ldots+u_n^3y_n$$

Numerical example:

x	1	4	7	10	13	
y	0	3	4	3	0	$[y]=10$
u	-2	-1	0	1	2	
uy	0	-3	0	3	0	$[uy]=0$
u^2y	0	3	0	3	0	$[uuy]=6$
u^3y	0	-3	0	3	0	$[uuuy]=0$

The second step involves the main computational work.

Third step

Linear regression:

$$y=a_0+a_1u$$

$$
\begin{aligned}
a_0 &= [y]/n \\
a_1 &= [uy]/S_2
\end{aligned}
$$

Quadratic regression:

$$y = b_0 + b_1 u + b_2 u^2$$

$$b_0 = \frac{[y]S_4 - [uuy]S_2}{n\,S_4 - S_2 S_2}$$

$$b_1 = a_1 \qquad\qquad \text{(see linear case)}$$

$$b_2 = \frac{[uuy]n - [y]S_2}{n\,S_4 - S_2 S_2}$$

Cubic regression:

$$y = c_0 + c_1 u + c_2 u^2 + c_3 u^3$$

$$c_0 = b_0 \qquad\qquad \text{(see quadratic case)}$$

$$c_1 = \frac{[uy]S_6 - [uuuy]S_4}{S_2 S_6 - S_4 S_4}$$

$$c_2 = b_2 \qquad\qquad \text{(see quadratic case)}$$

$$c_3 = \frac{[uuuy]S_2 - [uy]S_4}{S_2 S_6 - S_4 S_4}$$

S_2, S_4, and S_6 are sums which are tabulated in Table I of the Appendix (mathematical background: see also Appendix). All regression equations relate to a variable u rather than to the variable x. To calculate u for arbitrary values of x and vice versa, the following formulae are to be used:

$$a = (x_1 + x_n)/2 \qquad h: \text{ spacing}$$

n odd: $\quad u = (x - a)/h \qquad x = hu + a$
n even: $\quad u = 2(x - a)/h \qquad x = \tfrac{1}{2}hu + a$

When the coefficients of one of the formulae

$$y = a_0 + a_1 u$$
$$y = b_0 + b_1 u + b_2 u^2$$
$$y = c_0 + c_1 u + c_2 u^2 + c_3 u^3$$

are known, y may be calculated for arbitrary values of u. However, the actual calculation of powers of u may result in large rounding errors. This can be avoided by using the "nested" formulae

$$y = b_0 + u(b_1 + ub_2)$$
$$y = c_0 + u(c_1 + u(c_2 + uc_3))$$

Example: The steps for calculating a cubic expression:

—multiply u by c_3,
—add c_2,
—multiply the result by u,
—add c_1,
—multiply the result by u,
—add c_0.

End of calculation: the result is y.

Warning: The coefficients of the regression equations, a_0, a_1, b_0, etc., are ratios of differences. Each difference must be calculated accurately, otherwise the coefficients may be meaningless. The high sensitivity to rounding is typical for all types of least-square fitting.

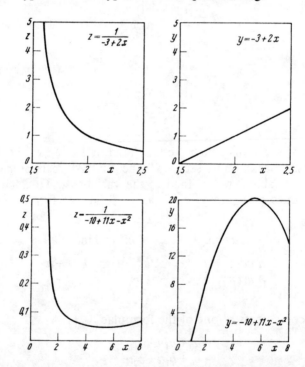

FIGURE 41. Hyperbolas and related curves. Outlined method also may be applied to hyperbolas "turned upside down," that is to convex curves.

Many nonlinear situations can be solved easily by using linear, quadratic, or cubic regresson. Three typical cases are considered next.

Case 1: The regression curve is a *hyperbola* or a closely related curve; for instance, the left-most graphs of Figure 41. The direct application of quadratic or cubic regression results usually in a poor fit of the data. The following simple approach yields in most instances excellent or at least acceptable results.

Let x and z be the variables. When plotting z on the ordinate versus the abscissa values x, the plot indicates a hyperbolical relationship between x and z. We calculate $y=1/z$, the reciprocal of z, and plot the new variable y versus x. If nearly a straightline results, linear regression should be applied to x and y. If, however, a concave or convex curve is indicated, quadratic or possibly cubic regression should be applied. Finally, if it seems that the plotted points are fitted best by a partly concave, partly convex curve, cubic regression is the answer. The calculated coefficients a_0, a_1, b_0, b_1, b_2, etc., refer to the following functions:

$$z=\frac{1}{a_0+a_1u}$$

$$z=\frac{1}{b_0+b_1u+b_2u^2}$$

$$z=\frac{1}{c_0+c_1u+c_2u^2+c_3u^3},$$

Case 2: *Exponential regression.* The regression curve is a straight-line if the points x,z are plotted on semilog paper whereby the x-values refer to the usual scaling of graph paper and the z-values to the logarithmic scale. Then the formula for the regression line is

$$y=a_0+a_1x \quad \text{where } y=\log z$$

The relation $y=\log z$ implies $z=10^y$, that is z is equal to ten to the power of a_0+a_1x. One also may write

$$y=\bar{a}+\bar{a}_1x \quad \text{where } y=ln\ z$$

is the natural logarithm of z, that is the logarithm to the base $e=2.71828\ldots$ Then

$$z=\exp(\bar{a}_0+\bar{a}_1x)$$

where $\exp(u)$ is the exponential function of the variable u.

Case 3: *Power laws.* A linear regression may result when plotting the variables w and z on double logarithmic paper. Then

$y = a + bx$ where $y = \log z$ and $x = \log w$.

This corresponds to the relation

$$z = 10^a w^b$$

(z is equal to ten to the power of the regression coefficient a *times* w to the power of the second coefficient b.)

Example 14.1   Correlation of the suspended material transported by a river with its flow velocity

Figure 42 refers to the Alpine part of the Rhine River. For details see Müller and Förstner (1968).

Example 14.2   A simple formula for calculating Student-t values

As we know, the Student-t values depend on the confidence level and the so-called degrees of freedom, v, which range from one to infinity (cf. Chapter 10). We now shall derive a simple formula to calculate t at the confidence level 90% for any specified value of v, that is for any degree of freedom.

First, the new variable $z = 1/v$ will be introduced. z ranges from 1 ($v = 1$) to 0 ($v = \infty$). When plotting t versus z on graph paper, a slightly curved line will result in the range from $z = 1/2$ ($v = 2$) to $z = 0$ ($v = \infty$). Thus, there should be a simple formula which covers the entire range with the possible exception of $z = v = 1$.

v	$t_{90\%}$	v	$t_{90\%}$	v	$t_{90\%}$
1	6.314	11	1.796	21	1.721
2	2.920	12	1.782	22	1.717
3	2.353	13	1.771	23	1.714
4	2.132	14	1.761	24	1.711
5	2.015	15	1.753	25	1.708
6	1.943	16	1.746	26	1.706
7	1.895	17	1.740	27	1.703
8	1.860	18	1.734	28	1.701
9	1.833	19	1.729	29	1.699
10	1.812	20	1.725	30	1.697
				40	1.684
				60	1.671
				120	1.658
				∞	1.645

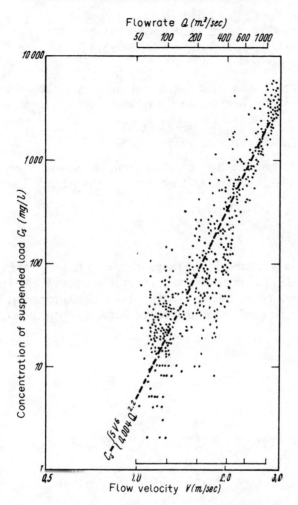

FIGURE 42. Concentration of suspended material (milligram per liter) versus flowing velocity of Alpine Rhine near its mouth at Lake "Bodensee." Figure represents slightly simplified version of original publication.

We select a quadratic interpolation formula

$$z = a + bz + cz^2$$

which has to satisfy the tabulated t-values for

$$z = 0 \qquad (v = \infty)$$
$$z = 1/3 \qquad (v = 3)$$
$$z = 1/5 \qquad (v = 5)$$

($z = 1/5$ is about one-half way between $z = 0$ and $z = \frac{1}{2}$)

$z=0$ $(v=\infty)$: $t=1.645=a$
$z=1/3$ $(v=3)$: $t=2.353=1.645+(b/3)+(c/9)$
$z=1/5$ $(v=5)$: $t=2.015=1.645+(b/5)+(c/25)$

The equation at the second line is multiplied by 9, the last equation is multiplied by 25. There results the system

$$c+3b=6.372$$
$$c+5b=9.250$$

When subtracting the first equation from the second, we will obtain $2b=2.878$, that is $b=1.439$. Hence $c=9.25-5b=2.055$. Therefore

$$t_{90\%}=1.645+1.439\,\frac{1}{v}+2.055\left(\frac{1}{v}\right)^{2}.$$

This formula is reliable in the range from $v=3$ to $v=\infty$; the maximum error is 0.003. The relative error always is less than 0.3%. The error is 1.4% for $v=2$ and intolerably large (about 100%) for $v=1$. Thus, the formula can be used for all degrees of freedoms with the exception of $v=1$.

CHAPTER 15

An Introduction to Discriminant Analysis

If two sample distributions overlap, three important statistical questions arise:

1. Do the samples belong to different populations?
2. If the first question is answered in the affirmative: To which population belong the individuals of the distribution overlapping region?
3. If the first question cannot be answered by some conventional test such as the Student-t test or the Chi-square method: Are there other ways to arrive at an answer?

Questions 2 and 3 are dealt with by a powerful method termed *discriminant analysis*. It can be applied if each sample is described by at least two numerical properties. This condition usually is satisfied. For instance, a clastic rock may be characterized by porosity, permeability, mean grain size, mineral composition, a sorting coefficient, etc. A fossil skull may be classified by using several metric parameters and so on.

The basic idea of discriminant analysis is simple. Let us compare two rock samples A and B. The distribution of each rock property for the samples may have a considerable overlap. For instance, in general, the grains of sample A may be larger than the grains of sample B, but the grain diameters x may overlap for too many grains, and a corresponding overlapping also may hold for some second property y. Because neither property x nor property y is sufficient to justify classifying the samples in different classes, we consider a *composite property X*, which is defined by the formula

$$X = ax + by$$

where a and b are some constant numbers. In general, the new composite property, X, is composed of completely different qualities. By selecting certain values for the constants a and b, the value of the composite property X can be calculated for each grain of the sample. There results two frequency distributions, one X-distribution for sample A, and one X-distribution for sample B. In general, these distributions also will

115

overlap. However, by selecting the numbers a and b properly, overlapping will be minimized. We then can decide whether the samples A and B belong to different populations and which items or individual of the samples belong to one population and which belong to another. In the most general situation, a linear combination $X = ax + by + cz + \ldots$ of n numerical properties $x, y, z, \ldots$ will be considered to classify items into different classes. The linear combination is termed a *discriminant function*. Obviously, the approach is useful, for instance in petrographic problems as well as in taxonomic problems of classifying individuals into various species or subspecies.

The combination of two properties

We compare a sample A of size n with a sample B of size m. There are two properties, x and y. Let x_{Ai} be the numerical value of property x, measured at the individual i of sample A. The symbols x_{Bi}, y_{Aj}, and y_{Bj} are defined correspondingly. For instance, y_{Bj} is the value of property y for individual j at sample B.

Numerical example:

| | Sample A | | | Sample B | |
i	x_{Ai}	y_{Ai}	j	x_{Bj}	y_{Bj}
1	172	130	1	191	128
2	180	124	2	187	112
3	184	130	3	180	117
4	168	124			
$n=4$			$m=3$		

The means: $\bar{x}_A = 176$ $\bar{x}_B = 186$
$\quad\quad\quad\quad\;\; \bar{y}_A = 127$ $\bar{y}_B = 119$

The ranges overlap: sample A: x: 168–184 y: 124–130
                     sample B: x: 180–191 y: 112–128

We now introduce the composite property $X = x - y$, that is $X = ax + by$ where $a=1$ and $b=-1$:

$$X = x - y$$

| | Sample A | | Sample B |
i	X_A	j	X_B
1	42	1	63
2	56	2	75
3	54	3	63
4	44		

The means: $\bar{X}_A = 49$ $\bar{X}_B = 67$

There is no overlapping; all sample values are separated.

Now we have to determine a general procedure for determining the constants a and b such that overlapping of the property $X = ax + by$ is minimized.

Obviously this is the situation if the distance between the means $\bar{X}_A$ and $\bar{X}_B$ is as large as possible and if the sum of the variances of X_A and X_B is as small as possible. To determine the optimum, we calculate

$$D_x = \bar{x}_A - \bar{x}_B \qquad D_y = \bar{y}_A - \bar{y}_B$$
$$u_{Ai} = x_{Ai} - \bar{x}_A \qquad v_{Ai} = y_{Ai} - \bar{y}_A \qquad i = 1,2,3,\ldots,n$$
$$u_{Bj} = x_{Bj} - \bar{x}_B \qquad v_{Bj} = y_{Bj} - \bar{y}_B, \qquad j = 1,2,3,\ldots,m$$

Then we determine the product sums

$$S_{11} = [u_A u_A] + [u_B u_B]$$
$$S_{12} = [u_A v_A] + [u_B v_B]$$
$$S_{22} = [v_A v_A] + [v_B v_B]$$

where

$$[u_A u_A] = u_{A1} u_{A1} + u_{A2} u_{A2} + u_{A3} u_{A3} + \ldots + u_{An} u_{An}$$
$$[u_A v_A] = u_{A1} v_{A1} + u_{A2} v_{A2} + u_{A3} v_{A3} + \ldots + u_{An} v_{An}.$$

etc.

Finally we calculate

$$b = (D_y S_{11} - D_x S_{12})/(D_x S_{22} - D_y S_{12}).$$

Optimum solutions are:

$$X = x + by \text{ and } X = -x - by$$

(It can be shown that there are an infinite number of equivalent optimum solutions. In other words, there are infinite number of optimum composite properties.)

Numerical example (continuation of the example):

i	u_{Ai}	v_{Ai}	$u_{Ai}u_{Ai}$	$u_{Ai}v_{Ai}$	$v_{Ai}v_{Ai}$
1	-4	$+3$	16	-12	9
2	$+4$	-3	16	-12	9
3	$+8$	$+3$	64	$+24$	9
4	-8	-3	64	$+24$	9
Sums:			160	24	36

j	u_{Bj}	v_{Bj}	$u_{Bj}u_{Bj}$	$u_{Bj}v_{Bj}$	$v_{Bj}v_{Bj}$
1	+5	+9	25	+45	81
2	+1	−7	1	−7	49
3	−6	−2	36	+12	4
Sum:			62	40	134

$$S_{11} = 160 + 62 = 222$$
$$S_{12} = 24 + 40 = 64$$
$$S_{22} = 36 + 134 = 170$$
$$D_x = 176 - 186 = -10$$
$$D_y = 127 - 119 = 8$$
$$b = -\frac{2416}{2212} = -1.09$$

Hence $X = x - 1.09y$ or $X = -x + 1.09y$. The first formula has the advantage that all X-values of the samples A and B are positive.

The combination of three properties

The formula for the composite property is

$$X = ax + by + cz.$$

We calculate D_x, D_y, S_{11}, S_{12}, and S_{22} as in the previous situation for two properties. Additionally we calculate

$$S_{13} = [u_A w_A] + [u_B w_B]$$
$$S_{23} = [v_A w_A] + [v_B w_B]$$
$$S_{33} = [w_A w_A] + [w_B w_B]$$
$$D_z = \bar{z}_A - \bar{z}_B$$

where

$$w_{Ai} = z_{Ai} - \bar{z}_A \qquad w_{Bj} = z_{Bj} - \bar{z}_B,$$

Finally, we calculate

$$b = \begin{vmatrix} S_{11} & D_x & S_{13} \\ S_{12} & D_y & S_{23} \\ S_{13} & D_z & S_{33} \end{vmatrix} \Big/ \begin{vmatrix} D_x & S_{12} & S_{13} \\ D_y & S_{22} & S_{23} \\ D_z & S_{23} & S_{33} \end{vmatrix}$$

$$c = \begin{vmatrix} S_{11} & S_{12} & D_x \\ S_{12} & S_{22} & D_y \\ S_{13} & S_{23} & D_z \end{vmatrix} \Big/ \begin{vmatrix} D_x & S_{12} & S_{13} \\ D_y & S_{22} & S_{23} \\ D_z & S_{23} & S_{33} \end{vmatrix}$$

Thereby the symbol

$$\begin{vmatrix} a_1 & b_1 & c_1 \\ a_2 & b_2 & c_2 \\ a_3 & b_3 & c_3 \end{vmatrix}$$

stands for the sum of products

$$a_1 b_2 c_3 + a_3 b_1 c_2 + a_2 b_3 c_1 - a_3 b_2 c_1 - a_1 b_3 c_2 - a_2 b_1 c_3.$$

Optimum solutions are

$$X = x + by + cz$$
$$X = -x - by - cz.$$

The combination of more than three properties

The simultaneous investigation of more than three properties is rather tedious and difficult without a computer program. In many situations, however, the investigation of certain combinations of two or three properties leads to results which are close to the optimum of the combination of all qualities.

CHAPTER 16

Introduction to the Construction of Geological Maps by Trend Analysis of Exploration Data*

The construction of geological maps which constitute sets of isolines such as lines of equal thickness, depth, etc., is one of the major tasks of the practicing geologist. From the mathematical point of view, the construction of isolines belongs to the art and science of interpolation: the isolines are to be constructed from values known at a few points on a geographical map. Strictly speaking, there is an infinity of solutions to this problem because every isoline constitutes an infinity of points. Actually, this lack of uniqueness is removed by the tacit application of rules which may not be realized by the practicing geologist and which reflect habits taught to him at university, personal experience, and prejudices believed to be experience. There usually is added a flavor of simplicity with the tendency for interpolating linearly between neighbouring points. As a consequence, there result geological maps which reflect the interpretation of incompletely known spatial distributions by an individual. Usually, maps of the same region drawn by different individuals will differ in detail but also may differ in fundamental respects such as the distribution of fault systems.

To save time, map construction by the geologist may be replaced by computer-oriented map construction. In general, this approach is no more accurate, more reliable, and less fanciful than the map of a geologist—it may even be inferior in some respects. A computer program however has the advantage that its interpretations rest on well-known rules of interpolation and trend analysis: the bias of the computer program is known. There is a second advantage, that may be more important: maps constructed by the same program are comparable, whereas maps constructed by different individuals differ in interpretation of the same facts. Thus, there is something to say in favor for

---

*This chapter is based on the publication of D. F. Merriam and J. E. Robinson, Trend analysis in geologic geophysical exploration (*Mathematical methods in geology and geophysics*: Hornicka Pribram vi vide a technic Pribram, Czechoslovakia, 1970).

standardized, sophisticated computer programs whose limitations are well known and taken into consideration.

Trend maps

A trend map displays the global features of a structure. All minor details are supressed. Small wrinkles and other local features are "smoothed-out." To construct a trend map, it usually is the most convenient approach to select a regression polynomial which gives a best fit of the known sampled values. Actually, this is a rather straightforward generalization of the method of least squares as explained in the chapter on correlation.

In the simplest situation, all *isolines* of the trend map are *straightlines* located on a *plane*. If the isolines refer to the position and depth of a layer, the plane of the trend map indicates the mean dip and strike of the formation. The mathematical equation of a plane is

$$z=A+Bx+Cy.$$

Thereby x and y are the coordinates of a point on the geographical map, and z is a sampled value of the considered stratum (as say depth or thickness or clastic/limestone ratio, etc.) at point x,y. The coefficients A, B, and C are constants which fix the plane. They are to be determined from known values x,y,z by the method of least squares. The equation of the isoline $z=z_0$ where z_0 is a fixed number as say $z_0=5000$ ft, is $Bx+Cy=z_0-A$. It can be constructed on the map by specifying two values of x and calculating the corresponding values of y from the last equation. The isoline is the straightline which passes the two points on the map.

To obtain more elaborate representations, a *quadratic surface*

$$z=A+Bx+Cy+Dx^2+Exy+Fy^2.$$

may be used. x, y, and z have the same representation as in the preceding example, the coefficients A to F determine the position and shape of the quadratic surface and must be calculated by the method of least squares. If $D=F=E=0$, the quadratic surface degenerates to the plane $z=A+Bx+Cy$. The equation $z=Fy^2$ represents a simple symmetrical valley whereby the coordinate system is oriented as shown in Figure 43.

Somewhat more interesting is the equation $z=Dx^2-Fy^2$ whereby the product DF must be positive. This equation represents a *saddle* which may be used as a model for a *mountain pass* (Figure 44). The point O, the origin of the coordinate system, that is $x=y=z=0$, is the highest point of the road *IOJ* which crosses the pass. All cuts parallel to *IOJ* are parabolas. All lines of equal height are hyperbolas with one exception: The lines passing through the origin O are two straightlines which

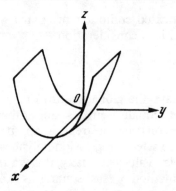

FIGURE 43. "Valley" with parabolic cross section generated by equation
$z = Fy^2$.

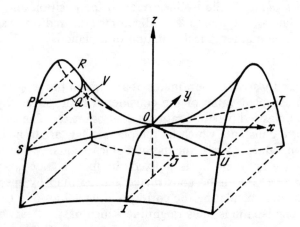

FIGURE 44. Quadratic surface forming saddle. For details see text.

intersect at O, that is at the highest point of the pass road. The intersecting lines are SOT and UOV (Figure 44).

Higher polynomials may be used to incorporate more detail. One should keep in mind, however, that no regression polynomial can be used for extrapolation to outside regions. Further, it is not recommended to use polynomials of degree five or larger. Polynomials of a high degree tend to generate nonexisting wavy surfaces.

Example 16.1   A trend-analysis map from Bohemia

Figure 45 shows a structural map from Bohemia. Figure 46 displays several trend maps on the structural map. Figure 46A shows the general strike and dip. The higher approximations reveal rather closely the general features of the structure.

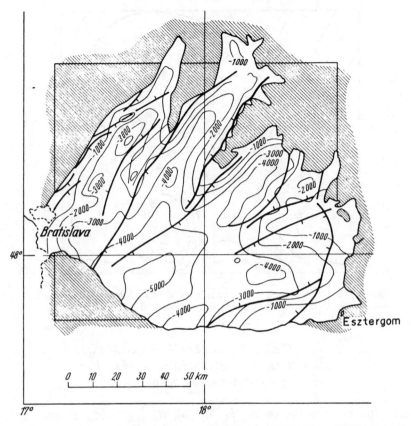

FIGURE 45. Structural map of Neogene basin between Bratislava (CSSR) and
Esztergom (Hungary).

Fourier representations

Another method to analyze structural maps is Fourier analysis. A
curve, time series, surface, or three-dimensional spatial distribution may
be represented by the superposition of sine and cosine curves. The
functions $z = \sin x$ and $z = \cos x$ represent regular waves of amplitude 2
and wavelength 2π ($\pi = 3.1417\ldots$ is equal to the circumference of a
circle divided by its diameter). The functions $z = A \sin nx$ and $z = A \cos$
nx describe waves of amplitude $2A$ and wavelength $2\pi/n$.

The function $z = \sin x + A \sin 3x$ represents the superposition of the
wave $\sin x$ and the wave $A \sin 3x$ (see Figure 47). More generally, a wave
train may be represented by the *trigonometric sum*

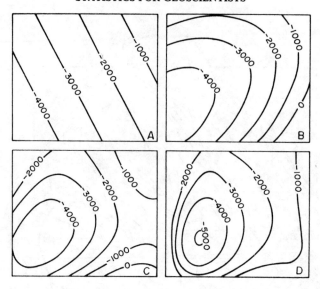

FIGURE 46. Trend maps corresponding to structural map of Figure 45.
A: linear trend, B: quadratic trend, C: cubic trend, D: quartic trend.

$$z = \alpha + a_1\cos(2\pi x/L) + b_1\sin(2\pi x/L) + a_2\cos(4\pi x/L) +$$
$$+ b_2\sin(4\pi x/L) + a_3\cos(6\pi x/L) + b_3\sin(6\pi x/L) +$$
$$+ \ldots + a_n\cos(2\pi n x/L) + b_n\sin(2\pi n x/L)$$

where the coefficients α, a_1, b_1, a_2, b_2, a_3, b_3, ..., a_n, b_n are to be determined by the method of least squares. The trigonometric sum represents $2n$ waves of the wavelengths L, $L/2$, $L/3$, $L/4$, ..., L/n. Actually, by properly selecting n and the coefficients α, $a_1, b_1 \ldots$, almost all curves of practical interest can be represented by trigonometric sums. Surprisingly, trigonometric sums even allow for the representation of jumps. As a consequence they may be used to *represent faulted structures*. A simple but typical sample is shown in the Figure 48; the approximation of a step curve by trigonometric sums composed of only six sine terms. To reduce the amplitude and wavelength of the small wrinkles, more than six terms must be used. Then, however, a slight "overshooting" occurs at the jump. This overshooting is known as "Gibbs phenomenon" because it was first noticed by the famous American physicist J. Willard Gibbs.

The trigonometric sums must be generalized slightly if surfaces rather than curves are to be represented. Let L be the longest wavelength in the x-direction of the geographical map, and let H be the longest wavelength in the map's y-direction (the x-axis may be selected arbitrarily; the y-axis is perpendicular to the x-axis). Then the following trigonometric double

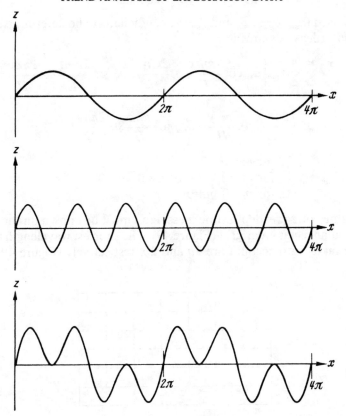

FIGURE 47. *Top*: Wave $z = \sin x$. *Middle*: Wave $z = \sin 3x$. *Bottom*: Superposition of waves $\sin x$ and $\sin 3x$.

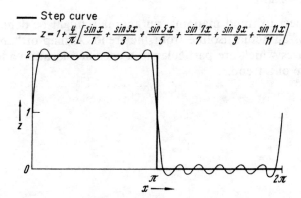

FIGURE 48. Approximating step curve by sum of sine terms.

sum is used (a_{mn}, b_{mn}, c_{mn}, and d_{mn} are coefficients to be determined by the method of least squares):

$$z = \sum_{m=0}^{M} \sum_{n=0}^{N} \lambda_{mn} \left[a_{mn} \cos\frac{2\pi mx}{L}\cos\frac{2\pi ny}{H} + b_{mn}\sin\frac{2\pi mx}{L}\cos\frac{2\pi ny}{H} + \right.$$

$$\left. + c_{mn}\cos\frac{2\pi mx}{L}\sin\frac{2\pi ny}{H} + d_{mn}\sin\frac{2\pi mx}{L}\sin\frac{2\pi ny}{H} \right]$$

$$\lambda_{mn} = \begin{cases} \frac{1}{4}: & \text{for } m=n=0 \\ \frac{1}{2}: & \text{for } m=0,\ n\neq0 \text{ or } n=0,\ m\neq0 \\ 1: & \text{for } m\neq0,\ n\neq0 \end{cases}$$

The values of z (z = depth, thickness, etc.) must be known on the points of a regular rectangular grid (Figure 49). The shortest wavelengths which can be taken into account are $2a$ and $2b$, respectively (Figure 49).

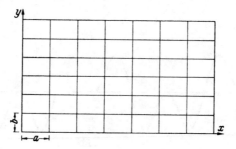

FIGURE 49. Rectangular grid.

Mapping "anomalies"

With respect to trend mapping, an anomaly might be defined as the deviation of an actual isoline from the trend. A map of "anomalies" shows features which are particular to a specified location and not the consequence of a trend.

CHAPTER 17

The Representation of Distribution Curves by Cubic Splines

When using the classical method of curve fitting by polynomials, two specified points are joined by a straight line with the equation $y=a_0+a_1x$, three points are fitted by a quadratic parabola $y=a_0+a_1x+a_2x^2$, four points are fitted by a cubic and so on. In general, a curve fitting $n+1$ points, may be constructed by properly selecting coefficients a_0, a_1, a_2, a_3,...., a_n of the polynomial:

$$y=a_0+a_1x+a_2x^2+a_3x^3+ \ldots a_nx^n.$$

This approach, however, may be disadvantageous because it may result in unwanted oscillations of the interpolation curve. Consider, for instance, the graph at the top of Figure 50. There are eight specified points located on the x-axis and one point located on the y-axis. The 8th degree polynomial whose curve passes through all nine points represents a wave. However, we may prefer a presentation with a minimum of oscillations such as in the graphs 2 and 3 of Figure 50. In other words, one prefers to join several specified points by a line whose arc length is as small as possible. Physically, this can be achieved by using a flexible metal strip, termed a "spline." Numerically, the curvature of an interpolation curve can be minimized by selecting appropriate cubic polynomials termed "cubic splines." The general approach presupposes a somewhat elaborated computer program. There is, however a satisfactory simpler solution to the problem.

We consider three consecutive points with the abscissa values $u=0$ $u=1$, and $u=2$, and with the corresponding ordinate values $y=a$, $y=b$, and $y=c$:

u	y
0	a
1	b
2	c

The curve matching these three points should be smooth, of minimum length, and should have the specified slope d at $u=0$. (The slope of a

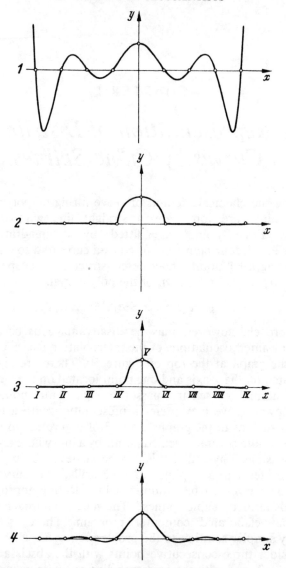

FIGURE 50. Interpolation by different types of polynomials. *Curve 1*: Classical interpolation. *Curve 2*: Interpolation curve is composed of two straight lines and arc. Curve has two corners. *Curve 3*: Interpolation by using method recommended in this chapter. *Curve 4*: Interpolation by conventional spline functions.

curve is the tangent of the curve's dip angle. For instance, a dip angle of 45° corresponds to a slope of $d=1$.) The curve of minimum or nearly minimum length which passes through the three points $(0,a)$, $(1,b)$, and $(2,c)$ with slope d at the first point is given by the equation

$$y = af_1 + bf_2 + cf_3 + df_4$$

where

$$f_1 = (3u^3 - 7u^2 + 4)/4$$
$$f_2 = -u^3 + 2u^2$$
$$f_3 = (u^3 - u^2)/4$$
$$f_4 = (u^3 - 3u^2 + 2u)/2$$

The slope e at the endpoint $u=2$, $y=c$ is equal to

$$e = 2a - 4b + 2c + d$$

(See T. J. Fletcher, *Linear Algebra*: Van Nostrand, London 1972, p. 13–14).

Example 17.1   Determine the spline function of the normal curve N(0,1), that is the Gaussian distribution curve with mean zero and variance 1 (Chapter 9.2.1). Calculate first the spline function for the range from x=0 to x=1

The value $x=0$ of the Gaussian curve corresponds to the value $u=0$ of the spline. The value $x=1$ corresponds to $u=2$. Hence, by using Table II of the Appendix, we obtain (in the table, the symbol z rather than x is used):

x	u	$N(0,1)$
0	0	$a=0.3989$
$\frac{1}{2}$	1	$b=0.3521$
1	2	$c=0.2420$

At $x=0$, the slope d of the normal curve $N(0,1)$ is zero. For at $x=0$, the dip angle of the curve is equal to zero because there the tangent is parallel to the x-axis. Thus, $d=0$.

Results:

x	$N(0,1)$	spline	u
0	0.3989	0.3989	0
0.1	0.3970	0.3968	
0.2	0.3910	0.3907	
0.3	0.3814	0.3810	
0.4	0.3683	0.3680	
0.5	0.3521	0.3521	1
0.6	0.3332	0.3337	
0.7	0.3123	0.3131	
0.8	0.2897	0.2907	
0.9	0.2661	0.2669	
1	0.2420	0.2420	2

The slope at $x=1$ is equal to $e=-0.1266$. The slope is negative because the curve is declining.

We now calculate the spline for the range $x=1$ to $x=2$:

x	u	$N(0,1)$
1	0	$a=0.2420$
1.5	1	$b=0.1295$
2	2	$c=0.0540$
		$d=-0.1266$

(The value of d of the second spline is equal to the value of e of the first spline because the endpoint of the first spline coincides with the first point of the second spline.)

Results:

x	$N(0,1)$	spline	u
1	0.2420	0.2420	0
1.25	0.1827	0.1820	
1.5	0.1295	0.1295	1
1.75	0.0863	0.0863	
2	0.0540	0.0540	2

The example indicates the general procedure. First, the curve is split into different parts. Then, for each part the splines are calculated and tabulated. All spline curves are arcs (or, in special situations, straight-lines) which form one single smooth curve. There are no corners at the joints of consecutive splines because they have the same slope at the point of intersection. For the first spline curve, one must know the value of the slope at $u=0$. All other required slopes are determined by the e-formula stated. The calculation of a nonvanishing first slope is shown in the next example.

Example 17.2 Determine spline functions of the cumulative Gaussian frequency distribution for positive values of x (Figure 21, Chapter 9.2.1)

First we consider the range from $x=0$ to $x=1.5$ (numerical values see Appendix Table II):

x	u	F (%)
0	0	$a=0$
0.75	1	$b=27.34$
1.5	2	$c=43.32$

The slope of the curve of F versus u is equal to the change dF of F divided by the corresponding change du of u:

$$slope = dF/du$$

To calculate d, that is the slope at $u=0$, we must read dF and du near to $u=0$:

x	F
0	0
0.05	2

$dx = 0.05 - 0 = 0.05$
$dF = 2 - 0 = 2$

$u = 1/15$ corresponds to $x = 0.05$ because $u = 1$ corresponds to $x = 0.75$: $(1/15)$: $1 = 0.05:0.75$. $dF = 2-0 = 2$ corresponds to $du = (1/15) - 0 = 1/15$. Consequently, the slope d at $u = 0$ is approximately equal to 2 divided by $1/15$, that is, $d = dF/du = 30$.

Results:

x	F	spline
0	0	0
0.25	9.87	9.82
0.5	19.15	19.04
0.75	27.34	27.34
1	34.13	34.38
1.25	39.44	39.81
1.5	43.32	43.32

$e = 7.28$

We now select a different x-range, namely 1 rather than 1.5:

x	u	F
1.5	0	$a = 43.32$
2	1	$b = 47.73$
2.5	2	$c = 49.38$
		$d = 7.28$ (the slope calculated previously)

Please note that the slope d at the starting point of the second spline must be equal to the slope $e = 7.28$ at the last point of the first spline to ensure smoothness at the point of intersection. We do not have to bother about the different x-ranges because the u-range always is the same. This is practical: at each step we can select different x-ranges, thereby always using the same e-formula.

Results:

x	u	F	spline
1.5	0.0	43.32	43.32
1.6	0.2	44.52	44.64
1.8	0.6	46.41	46.55
2.0	1.0	47.73	47.73
2.2	1.4	48.61	48.47
2.4	1.8	49.18	49.06
2.5	2.0	49.38	49.38

CHAPTER 18

Sampling from Inhomogeneous Universes

In the previous chapters it has been assumed tacitly that samples are drawn *at random*, favoring no individuals and no group of individuals. Otherwise, the sample statistics would be subject to systematic errors.

This is obvious. Nevertheless, it might be rather difficult to avoid nonrandom sampling. Locations which are important for random sampling of rock specimens or fossils may be inaccessible. It might be difficult to differentiate a sample composed of globular and needle-shaped particles into representative grain-size classes. An unexperienced statistician who tries to forecast a poll may question people from a single social stratum rather than representatives from all strata.

In any situation, nonrandom sampling may be difficult if a universe is a blend of different basic populations. Such a universe is termed "*inhomogeneous.*" For instance, two sands originating from different regions are blended to an "inhomogeneous" population when they are mixed, say, at the mouth of a river.

To distinguish among different types of inhomogeneous distributions, we may use the following classification:

1. *Homogeneous distributions.* The investigated property differs at all known locations in about the same manner. Most samples from different locations do not differ significantly. The relative number of significantly differing samples is in accord with statistical expectation.

2. *Inhomogeneous distributions.* The investigated property displays large variations. The geological body seems to be built-up of at least two homogeneous distributions of possibly different genetic origin. Samples from different locations differ significantly. There are essentially two limiting situations:

2.1. The different populations are intermingled randomly. Each universe is composed of isolated parts. Each part is surrounded by individuals of a different universe. There may be a trend. For instance, a clean, well-sorted sand may gradually merge into a sandy clay.

2.2. The parts of the different populations are composed of rather regular or rather well-defined units such as beds.

In any situation, different populations may merge continuously or may be distinguishable clearly such as dark parts in an otherwise light-colored rock.

To test the inhomogeneity of a universe, the methods of Chapters 20 and 21 may be used. We also may apply a Chi-square test.

Testing inhomogeneity by a Chi-square test

There are M samples. The first sample has the size N_1 (N_1 individuals), the second the size N_2, etc. Each sample can be split into classes A, B, C, ..., K such as grain sizes, types of minerals, etc.

Class	A	B	C	...	K	Sum of rows
Sample 1	a_1	b_1	c_1	...	k_1	N_1
Sample 2	a_2	b_2	c_2	...	k_2	N_2
......	.	.	.	...	.	
Sample M	a_M	b_M	c_M	...	k_M	N_M
Sum of columns	a	b	c	...	k	N
Mean	$P_A = a/N$	$P_B = b/N$	$P_C = c/N$	...	$P_K = k/N$	

a_1 is the absolute frequency of individuals belonging to sample 1 and class A, c_2 is the absolute frequency referring to sample 2 and class C, etc. The test refers to the contingency tables

f-distribution				
a_1	b_1	c_1	...	k_1
a_2	b_2	c_2	...	k_2
.	.	.	...	.
a_M	b_M	c_M	...	k_M

v-distribution				
N_1P_A	N_1P_B	N_1P_C	...	N_1P_K
N_2P_A	N_2P_B	N_2P_C	...	N_2P_K
...	...	...	...	...
N_MP_A	N_MP_B	N_MP_C	...	N_MP_K

$$\chi s^2 = \frac{(a_1 - N_1P_A)^2}{N_1P_A} + \frac{(b_1 - N_1P_B)^2}{N_1P_B} + \ldots + \frac{(k_M - N_MP_K)^2}{N_MP_K}$$

$v = (M-1)(K-1)$
$K =$ number of classes.

In a more detailed analysis, pairs and triples of samples, etc., are compared. This may be used to determine the number of homogeneous distributions that are blended to a heterogeneous population (see also Cadigan, 1962 and Mosimann, 1965).

Characterization of inhomogeneous universes by parameters

Let there be M samples. For each, some specified parameter Z such as a mean, or a variance, or a range correlation, etc., is determined. We calculate the mean and the variance

$$\bar{Z}=(Z_1+Z_2+ \ldots +Z_M)/M$$

and

$$\text{Var } Z=\{(Z_1-\bar{Z})^2+(Z_2-\bar{Z})^2+ \ldots +(Z_M-\bar{Z})^2\}/M.$$

Then Z is the best estimate of the parameter Z of the inhomogeneous population. The confidence limits of $\bar{Z}$ are approximately

$$\bar{Z}\pm tA$$

where $A=$ square root of (Var $Z)/(M-1)$. The value of t has to be read from Table III of the Appendix for $M-1$ degrees of freedom. Because $\bar{Z}$ is a mean value, two values of $\bar{Z}$ may be compared by applying the Student-t test. If they do not differ significantly, they may be attributed to the same heterogeneous universe.

Example 18.1   Heterogeneous distributions of sands

Samples taken from a sand contain usually large numbers of grains. As a consequence, grain-size sample distributions should be nearly identical and close to the grain-size distribution of the whole sand. Actually, however, in many instances this does not hold because many sands constitute mixed populations.

Example 18.2   Heterogeneous distributions in climatology

Climatic events usually follow a pattern which may be described as partly regular and partly random. For instance, there is a tendency of the accumulation of rainy (dry) days or rainy (dry) years. This is shown by measurements of water-depth gauge in African, North American, and European rivers; in the Nile for more than a thousand years (for details see Leopold, Wolman, and Miller, 1964, p. 56–58).

CHAPTER 19

Sampling Methods. Monte Carlo Sampling

In many instances, actual sampling in the geosciences can not be considered as the ideal random sampling required by mathematical statistics. Actual sampling might be restricted to a few locations or hampered by lack of time or lack of financial resources. In the following, it will be assumed that samples can be collected at numerous locations which cover sufficiently the investigated geological body (or profile, layer, etc.). Then there are essentially six sampling methods available.

1. Systematic sampling

Samples are collected at regular intervals. This is a possibly time-consuming but excellent method for a homogeneous universe. On the other hand, systematic sampling results in biased estimates if the investigated property changes periodically with a period corresponding to the sampling interval.

2. Random sampling

In general, by random sampling a bias is avoided. A map of the geological body is drawn up and covered by a regular grid. Each grid point is numbered according to some regular scheme. Then *random numbers* are drawn (say from a table of random numbers such as Table VII of the Appendix) and samples of size N are collected at each of those M grid points whose marks coincide with a drawn random number.

Example. There are 87 grid points numbered from 01 to 87. Ten random numbers are drawn one of which occurs twice:

16, 54, 47, 17, 09, 35, 61, 77, 73, 16.

Hence, a total of $M=9$ samples of size N are drawn, exactly one from the following locations:

09, 16, 17, 35, 47, 54, 61, 73, and 77.

The reliability of the data is checked by a *variance test*. For each sample of size N a parameter Z is determined where Z may be the sample mean $\bar{x}$, or the sample variance s^2, or s, or one of the sample quartiles Q_1, Q_2, Q_3. There result M numbers Z_1, Z_2, ..., Z_M. We then calculate

$$\bar{Z}=(Z_1+Z_2+\ldots+Z_M)/M$$
$$\text{Var } Z=\{(Z_1-\bar{Z})^2+(Z_2-\bar{Z})^2+\ldots+(Z_M-\bar{Z})^2\}/M$$

and compare Var Z with the sample which is composed of all M collected samples.

Example. There are $M=9$ samples of size $N=50$ each, that is each sample consists of fifty numbers representing the sampled variability of some property. Z is selected as the sample mean $\bar{x}$. There result nine values $\bar{x}_1$, $\bar{x}_2$, ...,$\bar{x}_9$ and a corresponding value of Var $\bar{x}$. The nine samples are put together to form a new single sample of size $M\times N=9\times 50=450$. With other words, the new sample consists of the 450 numbers sampled at nine locations.

We now calculate the variance σ^2 of the new sample and compare $\hat{s}_1^2=$Var Z with $\hat{s}_2^2$ (see the following table) by applying Snedecor's F-test of Chapter 11.1. We use Table IV of the Appendix with $M-1$ degrees of freedom for $\hat{s}_1^2$ and $NM-1$ degrees of freedom for $\hat{s}_2^2$. The sampling can be considered as unbiased if $\hat{s}_1^2$ and $\hat{s}_2^2$ do not differ significantly.

$\hat{s}_2^2$ of the variance test for random sampling

Z	$\hat{s}_2^2$	Remarks
x	σ^2/N	$N \geq 30$
Q_2	$1.57\sigma^2/N$	$N\geq 30$
Q_1,Q_3	$1.86\sigma^2/N$	$N\geq 30$
s	$\sigma^2/2N$	$N\geq 100$
s^2	$2\sigma^4/N$	$N\geq 100$

This table refers to the situation that the universe is very large and that the sample size N is at most one tenth of the sample size G of the universe. For finite universes, the exact formula for $Z=\bar{x}$ is known:

$$\hat{s}_2^2=\sigma^2\left\{\frac{1}{N}-\frac{1}{G}\right\}$$

In nearly all applications of statistics to the geosciences, the size G of the universe is so large that $1/G$ can be ignored safely.

Example. There are five samples of size 30 each. The sample mean is specified as parameter Z. Consequently

$$M = 5$$
$$N = 30$$
$$Z = \bar{x}$$
$$\hat{s}_1^2 = \text{Var } \bar{x} \text{ with 4 degrees of freedom}$$
$$\hat{s}_2^2 = \sigma^2/30 \text{ with 149 degrees of freedom}$$

Please keep in mind that doubling of sample size N results in halving the variance.

If the variance test indicates the reliability of the performed sampling, then confidence limits of $\bar{Z}$ may be determined by applying the Student-t test. The confidence limits are

$$\bar{Z} \pm t \text{ times square root of (Var } Z)/(M-1).$$

The variance test also may be used to investigate *time series*. Are the positive and negative deviations from trend randomly distributed or are they *clustered*?

3. Stratified sampling

This method is applied if the universe is heterogeneous but composed of k clearly distinguishable homogeneous universes ("strata") with known relative frequencies $q_1, q_2, q_3, \ldots, q_k$ $(q_1+q_2+q_3+ \ldots +q_k=1)$.

Example: The universe is a profile, q_1 is the mean relative thickness of the lowest layer.

One sample is collected from each stratum. Thus, there are k samples with corresponding means (or variances or quartiles etc.) $Z_1, Z_2, Z_3, \ldots, Z_k$. Then the best estimate of the Z-value of the total heterogeneous universe is the weighted mean of Z-values

$$Z=q_1Z_1+q_2Z_2+q_3Z_3+ \ldots +q_kZ_k$$

If Z is the mean $\bar{x}$ of the total universe, the variance of $\bar{x}$ can be estimated by the formula

$$\text{Var } \bar{x}=(q_1^2\hat{s}_1^2/N_1)+(q_2^2\hat{s}_2^2/N_2)+ \ldots +(q_k^2\hat{s}_k^2/N_k)$$

$N_1, N_2, N_3, \ldots, N_k$ are the sample sizes; $\hat{s}_1^2, \hat{s}_2^2, \ldots$ are the sample variances (best estimates). To keep Var $\dot{x}$ small, the sample size must be large if both the relative frequency of the corresponding stratum and the corresponding sample variance are large.

4. Cluster sampling (nested sampling)

The collection of samples is restricted to a few adjacent locations. This method cannot be recommended but may be dictated by circumstances.

The terms *subsampling* and *two-stage sampling* signify essentially that more than one sample is collected from selected parts of the universe.

5. Biased sampling

There always is the danger of biased sampling, and it might be difficult to escape it. A typical example is overestimating the volume of a rock's dark minerals.

The statistical aspects of sampling are considered for instance by Cochran (1966) and Krumbein and Graybill (1965), Chapters 7 and 9. Technical aspects are dealt with by Oelsner (1952).

6. Monte Carlo sampling

Monte Carlo sampling is the attempt to solve probabilistic problems by mathematical experiments (involving random numbers) rather than by rigorous mathematical reasoning. Involved problems requiring an exceptional degree of mathematical ingenuity can be handled by Monte Carlo sampling easily and conveniently on a strictly elementary level. A typical example is the optimization of a complex communication system. The duration, number and temporal distribution of calls are determined by statistical frequency distributions. How many lines must be installed such that the system is neither jammed nor idle for an uneconomical span of time?

Example 19.1   The irregular chopping-off of a part of a grain-size distribution

The following example of Monte Carlo sampling is hypothetical and fairly simple but requires some attention on the part of an unexperienced reader. The example may be skipped by anyone not interested in the investigation of complex, mutually influencing geostatistical phenomena.

A continuously effective event A produces currently a grain-size distribution S_A which changes with time. Additionally, there is a second event B of irregularly changing intensity chopping off parts of S_A. Thus, B is a type of statistical filter which withholds parts of S_A. What is the final cumulative frequency distribution of grain size?

Some properties of event B:

Intensities:	I	II	III
Frequency:	25%	60%	15%
Cumulative frequency:	25%	85%	100%
Chops off all grains larger than	6	14	25

The grain-size distribution S_A is composed actually of three different distributions S_1, S_2, and S_3 which occur with frequencies q_1, q_2, and q_3, respectively. The occurrences are in an irregular temporal sequence. However, the numerical values of q_1, q_2, and q_3 depend on the intensity of event B:

	q_1	$q_1 + q_2$	$q_1 + q_2 + q_3$
I	14	29	100%
II	24	58	100%
III	37	69	100%

Solution of the problem. A random number is drawn, say 49. This number is interpreted as a cumulative frequency related to the intensity of B. Because 49 is smaller than 85 but larger than 25, it refers to intensity II which chops off all grains larger than 14. Next a q-frequency is determined. A random number is drawn, say 18. For intensity II, it relates to q_1. Hence $S_A = S_1$. The first numerical experiment results in the final grain-size distribution S_1 chopped off at grain size 14. This cumulative distribution has to be readjusted to make up to 100% at the grain size 14. Then a second numerical experiment is performed resulting in an additional chopped-off distribution. It also is readjusted to 100% and added to the first distribution to obtain the average distribution of both. This process is repeated until the merged distributions stabilize. The final distribution is the grain-size distribution generated by the combined, interdependent causes A and B.

CHAPTER 20

Introduction to the Analysis of Variance (ANOVA)

The analysis of variance is a powerful tool for comparing any number of samples simultaneously. Strictly speaking, the method is restricted to sampling from a normal universe. For instance, in the situation of a lognormal population, the values of log x rather than the values of the variable x should be analyzed.

The gist of the method

When collecting data at random from a homogeneous population, we should not expect each sample to have the same mean, because even sample averages must reflect the variance in the parent population. We, however, should expect that the variation between samples is not larger significantly than the variation of data within each single sample. We will suspect that different samples were drawn from different universes if the "between sample variation" is distinctly larger than the "within sample variation." In this situation, we may be concerned with a heterogeneous universe that is composed of several homogeneous populations.

The calculation routine

Let there be k samples of data from a variable x with sample sizes n_1, n_2, $n_3, \ldots, n_k$. First, all items are collected into one "grand sample" comprising all samples:

Total number of x-values:	$N = n_1 + n_2 + \ldots + n_k$
Sum of all x-values:	$T = x_1 + x_2 + \ldots + x_N$
Mean of grand sample:	$\bar{x} = T/N$
Sum of squares of grand sample:	$GQS = (x_1 - \bar{x})^2 + (x_2 - \bar{x})^2 + \ldots + (x_N - \bar{x})^2$
Variance of grand sample:	$GQS/(N-1)$
Degrees of freedom:	$N-1$

Secondly, the variance between samples must be calculated. To do so, we must eliminate the variability within samples be replacing each x-value by its corresponding mean. The following formulae result ($\bar{x}_1$, $\bar{x}_2$, $\bar{x}_3, \ldots, \bar{x}_k$ are the means of the k collected samples):

Square sum of means: $\quad ZQS = n_1(\bar{x}_1 - \bar{x})^2 + n_2(\bar{x}_2 - \bar{x})^2 + \ldots + n_k(\bar{x}_k - \bar{x})^2$
Between sample variance: $\quad ZQS/(k-1)$
Degrees of freedom: $\quad k-1$

Thirdly, the "within sample variance" must be calculated. It can be shown that the corresponding square sum is equal to GQS$-$ZQS. The number of freedoms are $(N-1)-(k-1)=N-k$. Thus, the "within sample variance" is equal to $(GQS-ZQS)/(N-k)$. Finally, the "within sample variance" and the "between sample variance" are to be compared by Snedecor's F-test.

Example 20.1 Numerical example

Sample 1, listed values: 2, 3, 1 $n_1 = 3$ $\bar{x}_1 = 2$
Sample 2, listed values: 2, 3, 1, 1, 3 $n_2 = 5$ $\bar{x}_2 = 2$ $\left.\right\}$ $k = 3$ $N = 12$
Sample 3, listed values: 2, 4, 8, 6 $n_3 = 4$ $\bar{x}_3 = 5$

$$\bar{x} = (2+3+1+2+3+1+1+3+2+4+8+6)/(3+5+4) = 3$$
$$GQS = (2-3)^2+(3-3)^2+(1-3)^2+(2-3)^2+(3-3)^2+(1-3)^2+$$
$$(1-3)^2+(3-3)^2+(2-3)^2+(4-3)^2+(8-3)^2+(6-3)^2 = 50$$
$$ZQS = 3(2-3)^2+5(2-3)^2+4(5-3)^2 = 24$$

Variation	Sum of squares	Degrees of freedom	Variance
Total	50	$12-1=11$	$50/11$
"between"	24	$3-1=2$	$24/2 = 12$
"within"	$50-24=26$	$11-2=9$	$26/9 = 2.9$

Variance ratio: $12/2.9 = 4.1$
Snedecor's F-test (5%):   4.3   (interpolated)
Conclusion: Homogeneous population ($\pm$)

Example 20.2 Paleogeographic barriers

It is conjectured that two regions were separated in the past by some barrier. Samples are collected from different locations. For each region, the variation between samples is compared to variation within samples. Finally, the grand mean of the first region is compared to the grand mean of the second region by applying some test such as Student's-t test.

Example 20.3   Comparing recorded measurements of different observers

The analysis of variance may be used for comparing the observations of different scientists studying the same phenomenon. Do the results differ systematically?

Example 20.4   A slight generalization. Investigating a profile

The investigated profile consists of k beds. From each bed, m samples are collected at fixed horizontal distances to gather quantitative information about possible facies changes. (*Example*: samples are collected from the left-most corner of the profile at the horizontal distances 20 ft, 40 ft, 60 ft, etc.) We calculate the square sum of the k means between beds, ZQSB, the square sum of the m means between horizontal sampling locations, ZQSH, the grand total, GQS, and the corresponding variances with the degrees of freedoms $k-1$, $m-1$, and $N-1$ respectively ($N=mk$). For instance, the variance between beds is $ZQSB/(k-1)$. Do the variances differ between beds and horizontal locations significantly? Are there significant facies changes? Are there significant differences between different layers?

Finally we calculate the *residual variance* $(GQS-ZQSB-ZQSH)/M$ where $M=(N-1)-(k-1)-(m-1)$ is the number of freedoms of the rest variance. Is the residual variance negligible? The residual variance is a measure of a variation which cannot be reduced to facies variability or variation between different layers.

Simple problems of variance analysis are dealt with for example in the works of Curray and Griffiths (1955), Griffiths (1953), Krumbein and Miller (1953), and Krumbein (1954). For complex problems see for instance Miller and Kahn (1965, chap. 10 and 13), Simpson, Roe, and Lewontin (1960, chap. 12), and Snedecor (1948, chap. 10).

CHAPTER 21

Latin Squares

Nearly all phenomena depend causally or otherwise on several factors rather than on one. When studying the outcome of experiments as functions of different independent variables in the physical or chemical sciences, the investigation is split into as many series as there are number of variables (factors). For each series, all factors are kept constant except one which is changed.

This systematic approach usually can not be realized when studying a phenomenon in its natural surroundings or when performing experiments on a heterogeneous material. In this situation factor analysis (Chapter 22) or the method of latin squares may be helpful. Both approaches shall be developed step by step by means of samples of increasing complexity.

Example 21.1   The effect of different treatments on the yield of wheat

We wish to investigate the effect of three different treatments A, B, and C on the yield of wheat. We may subdivide a block of land into plots of equal area with three rows and three columns. To eliminate the effect of a possible systematic change of soil fertility, we apply each treatment to three plots out of the total of nine, the plots being selected at random such that each treatment occurs exactly once in each row and each column. Such a special type of a chess board-like array of n rows and n columns is termed a *latin square* arrangement.

Example of a 3×3 latin square for investigating the effect of three factors, A, B, and C:

A	B	C
B	C	A
C	A	B

A corresponding systematic arrangement of three strips:

A	B	C
A	B	C
A	B	C

The yields from the three strips may reflect changes in soil fertility rather than the effect of different treatments if soil fertility changes from east to west.

The calculation routine for a layout of latin squares

We wish to study the effect of n different causes (or n different intensities of the same cause) on the outcome of an event. The latin square consists of n rows and n columns with a total of $N=n^2$ boxes. Each box is filled with exactly one number representing the outcome of an experiment or another event. Thus, the latin square contains $N=n^2$ numbers $x_1, x_2, x_3, \ldots, x_N$. We calculate:

Sum of all x-values: $T = x_1 + x_2 + \ldots + x_N$

Square sum of grand sample: $GQS = x_1{}^2 + x_2{}^2 + \ldots + x_N{}^2 - \dfrac{T^2}{N}$

We sum the x-values of each *row*. There result numbers $y_1, y_2, y_3, \ldots, y_n$ corresponding to n rows (one number for each row). We calculate:

"Between-rows" square sum: $RQS = \dfrac{1}{n}(y_1^2 + y_2^2 + \ldots + y_n^2) - \dfrac{T^2}{N}$

We sum the x-values of each *column*. There result numbers $z_1, z_2, z_3, \ldots, z_n$ corresponding to n columns (one number for each column). We calculate:

"Between-column" square sum: $CQS = \dfrac{1}{n}(z_1^2 + z_2^2 + \ldots + z_n^2) - \dfrac{T^2}{N}$

We sum the x-values which correspond to the first factor, the second factor, etc. There result numbers $u_1, u_2, u_3, \ldots, u_n$ corresponding to n factors (one number for each factor). We calculate:

"Between-factors" square sum: $FQS = \dfrac{1}{n}(u_1^2 + u_2^2 + \ldots + u_n^2) - \dfrac{T^2}{N}$

146   STATISTICS FOR GEOSCIENTISTS

There results the following table:

Variability	Sum of squares	Degrees of freedom	Variance
Grand total	GQS	$N-1$	$GQS/(N-1)$
Between rows	RQS	$n-1$	$RQS/(n-1)$
Between columns	CQS	$n-1$	$CQS/(n-1)$
Between factors	FQS	$n-1$	$FQS/(n-1)$

Finally, we calculate the residual variability. The sum of squares of the rest is defined as

$$DQS = GQS - RQS - CQS - FQS$$

with the degrees of freedoms $N+2-3n$

$$[N+2-3n = N-1-3(n-1)].$$

Thus, the residual variance corresponding to the rest variability is equal to $DQS/(N+2-3n)$.

The variances between rows and between columns should not differ significantly, otherwise the outlay is not random. There is a significant effect of at least one factor if the variance between factors is larger significantly than the rest variance. In this situation, the effect of each factor may be investigated more closely by applying the Student-t test to each pair of factors.

Example 21.2   A latin square with 25 blocks

Some property of a plant is investigated in five regions A, B, C, D, and E. Is there some factor which is effective in all regions? The property is recorded at five different random spots of region A (measured values: α_1, α_2, α_3, α_4, α_5), at five spots of region B (values: β_1, β_2, β_3, β_4, β_5), etc. Then a latin square is set-up:

A α_1	B β_1	C γ_1	D δ_1	E ε_1
D δ_2	E ε_2	A α_2	B β_2	C γ_2
B β_3	C γ_3	D δ_3	E ε_3	A α_3
E ε_4	A α_4	B β_4	C γ_4	D δ_4
C γ_5	D δ_5	E ε_5	A α_5	B β_5

We calculate GQS, RQS, CQS, FQS, DQS, and the corresponding variances where FQS is the sum of squares between regions (thus regions are considered as factors.) There might be an additional factor if the rest variance is significantly larger than the variances between rows and between columns.

Example 21.3   Testing the homogeneity of a rock with respect to different rock properties

Sixteen cores are sampled. Four cores each are tested by four different methods A, B, C, and D (methods are considered as "factors"). For instance, method A may refer to porosity, method B to rock compressibility, etc. The results are filled in a 4×4 latin square. There are no significant local inhomogeneities if the rest variance does not differ significantly from the variances between rows and between columns.

The same test may be applied for comparing four different methods measuring the same petrophysical property.

Example 21.4   Investigating the effect of rock type and other factors on cementation

Recorded values (percentage of bulk volume):

Rock type	Cement (%)				Mean	Coded values=cement%−10			
A=greywacke	2	3	3	4	3	−8	−7	−7	−6
B=arkose	8	9	11	12	10	−2	−1	1	2
C=sandstone	8	11	12	17	12	−2	1	2	7
D=others	12	14	16	18	15	2	4	6	8

The variances do not change if all measured values are changed by adding a constant. To simplify the representation of the following calculations, all recorded measurements are diminished to values between −8 and +8 by subtracting 10 ("coded measurements"). We set-up a latin square and fill in the coded measurements by using random numbers. (Example: the first drawn random number is 2. The second coded number of rock type A (greywacke) is −7. We therefore fill-in the number −7 in the topmost cell of rock type A):

1) A -7	2) B -2	3) C -2	4) D 6	-5
5) D 2	6) A -8	7) B 1	8) C 2	-3
9) C 1	0) D 4	1) A -6	2) B -1	-2
3) B 2	4) C 7	5) D 8	6) A -7	10
-2 A -28	$+1$ B 0	$+1$ C 8	0 D 20	$T=\ 0$ $N=16$ $T^2/N=\ 0$

Variability	Sum of squares		Degrees of freedom	Variance
Grand total	$7^2+2^2+2^2+\ldots+8^2+7^2-0$	$=386$	$16-1=15$	25.7
Between rows	$\frac{1}{4}(5^2+3^2+2^2+10^2)-0$	$=\ 34.5$	$4-1=\ 3$	11.5
Between columns	$\frac{1}{4}(2^2+1^2+1^2+0^2)-0$	$=\ 1.5$	$4-1=\ 3$	0.5
Between factors	$\frac{1}{4}(28^2+0^2+8^2+20^2)-0$	$=312$	$15-3-3-3=\ 6$	104
Residual			38	6.3

The variance ratio between rows and columns is large and significant at the 5% level of Snedecor's F but not significant at the 1% level. The ratio between the rest variance and the variance between factors is significant at the 1% level. The first result indicates that we should repeat the calculations with a different, more randomly arranged, latin square. The second result indicates that rock type effects cementation. The residual variance is small. Thus, there seems to be no relevant factor which is independent of rock type.

CHAPTER 22

Factor Analysis

If a phenomenon is affected by several factors, the combined affect may be the result of adding the intensities of all factors. There also might be an interaction between factors resulting in a mutual weakening of each factor. For instance, a biological growth parameter may be large in spring and large at some locality named A. As a consequence, the parameter may become very large at A during spring. It also may be the situation, however, that the parameter is limited by a maximum value which always is reached at locality A. In this situation, spring has no affect at A.

The analysis of interaction presented here rests on the analysis of variance. The more recent approaches are beyond the scope of this book.

Example 22.1   Analyzing diagenetic processes

The porosity of a diagenetically changed sandstone was measured at different localities. Part of the pore space is filled with water, part with gas. Three hypotheses are to be tested (Füchtbauer, 1967):

1. Pore content is a determining factor of porosity because gas filling interrupts diagenenis such as cementation (symbols: I_1: water filling, I_2: gas filling).
2. The distance from the boundary of the sandstone bank is a factor because cementation seems to be most pronounced at the border zone (symbols: a_1: distance smaller than 5 ft, a_2: distance larger than 5 ft).
3. The main cementation material, anhydrite, is mainly in zones of relatively large grain sizes (symbols: k_1: median of grain-size distribution smaller than 0.2 mm, k_1: median larger than 0.2 mm).

149

Table of measured porosities (percentage values minus 10)

I_1				I_2			
a_1		a_2		a_1		a_2	
k_1	k_2	k_1	k_2	k_1	k_2	k_1	k_2
-8	-7	-7	-4	-8	-7	1	-3
-8	-6	-1	-4	-8	-6	4	1
-7	-6	4	-2	-7	-5	4	2
-7	-4	4	-2	-6	-5	4	2
-4	-3	6	-1	-3	-3	4	6
-3	-3	7	-1	-3	-3	8	7
-3	-2	7	0	-1	-2	8	7
-2	-2	7	1	3	-1	13	8

We calculate the mean and the variance for each combination of factors:

	$\bar{x}$	s^2		$\bar{x}$	s^2
$I_1a_1k_1$	4.75	5.44	$I_1a_2k_2$	8.37	2.82
$I_1a_1k_2$	5.87	3.42	$I_1a_2k_1$	13.37	21.87
$I_2a_1k_1$	5.87	13.17	$I_2a_2k_2$	13.75	12.94
$I_2a_1k_2$	6.00	3.75	$I_2a_2k_1$	15.75	12.19

Combinations with the factor a_1 (border zone): The Student-t test shows that there is no significant difference between the smallest (4.75) and the largest (6.00) mean. Hence there is no measurable affect of pore contents and grain size on diagenesis at the border zone.

Combinations with the factor a_2 (interior zone): The means of the a_2-combinations differ significantly or probably significant from the largest mean of the a_1-group. Thus, the effect of distance from the boundary of the bank is pronounced.

The means and variances of the combinations $I_1a_2k_1$, $I_2a_2k_2$, and $I_2a_2k_1$ do not differ significantly. The variances of $I_1a_2k_2$ and $I_1a_2k_1$ differ significantly. Thus, the Student-t test of Chapter 11.2 cannot be applied for comparing $I_1a_2k_2$ and $I_1a_2k_1$. We therefore calculate the confidence limits for each mean according to Chapter 10, determining at the 5% level:

$$I_1a_2k_2 \text{ (mean: } 8.37): \qquad 6.85 \leq \bar{x} \leq 9.89$$
$$I_1a_2k_1 \text{ (mean: } 13.37): \qquad 9.13 \leq \bar{x} \leq 17.61$$

These intervals overlap slightly. Obviously, this result is difficult to interpret and due to the high variance of the combination $I_1a_2k_1$. By continuing in this way we determine finally:

Distance: a_2, pore contents: differs

Small grain size, different filling (comparing I_1k_1 with I_2k_1):
  the means do not differ significantly

Larger grain size, different filling (comparing I_1k_2 with I_2k_2):
  the means differ significantly or probably significant

Result: Pore contents affects porosity in the parts which are remote from the border zone and composed of a rather coarse material (median grain size larger than 0.2 mm). Pore contents does not affect porosity in the border zone and in regions where the grain sizes are small. It is not clear whether the last result is due to physical causes or to nonrepresentative sampling.

Variation of grain size: When comparing I_1k_1 with I_1k_2 (water filling) and I_2k_1 with I_2k_2 (gas filling), it turns out that the means do not differ significantly. Hence, the median does not seem to effect porosity.

Now variance analysis will be applied to show:

1. The effect of the distance from the border zone is highly significant.
2. The effect of pore contents is significant.
3. There is no recognizable effect of grain size.

First step of analysis: Each factor is considered separately. All values corresponding to different classes of I, a, and k respectively are added:

I		a		k	
I_1	I_2	a_1	a_2	k_1	k_2
-61	$+11$	-140	$+90$	-2	-48

We calculate:

T =sum of all values
N =number of all values (total sample size)
N_c =number of classes of factor c
T $=-61+11=-140+90=-2-48=-50$
T^2/N $=2500/64=39$
N/N_c $=64/2=32$

(In this example, each factor is split into two classes. In general, the number of classes are different for different factors.)

We calculate:

$$QS=\frac{\text{sum of squares}}{N/N_c}-\frac{T^2}{N}$$

with N_C-1 degrees of freedom:

QS_I $= \frac{1}{32}$ $(61^2+11^2)-39$ $= 81$
QS_a $= \frac{1}{32}$ $(140^2+90^2)-39$ $=827$
QS_k $= \frac{1}{32}$ $(2^2+48^2)-39$ $= 33$
degrees of freedom: $2-1=1$

Second step of analysis: Each pair of factors is considered.

$I \times a$				$I \times k$				$a \times k$			
$I_1 a_1$	$I_1 a_2$	$I_2 a_1$	$I_2 a_2$	$I_1 k_1$	$I_1 k_2$	$I_2 k_1$	$I_2 k_2$	$a_1 k_1$	$a_1 k_2$	$a_2 k_1$	$a_2 k_2$
-75	$+14$	-65	$+76$	-15	-46	$+13$	-2	-75	-65	$+73$	$+17$

$$T^2/N = 39 \qquad N/N_c = 64/4 = 16$$

(There are four different classes of factor combinations for each combination of two factors.)

We calculate (α and β are any pure factors):

$$QS_{\alpha \times \beta} = \frac{\text{sum of squares}}{N/N_c} - \frac{T^2}{N} - QS_\alpha - QS_\beta$$

with

$$v_{\alpha \times \beta} = v_\alpha v_\beta$$

degrees of freedom (v_α is the number of degrees of freedom of QS_α):

$$QS_{I \times a} = \tfrac{1}{16}(75^2 + 14^2 + 65^2 \ 76^2) - 39 - 81 - 827 = 42 \qquad v_{I \times a} = 1$$
$$QS_{I \times k} = 4 \quad v_{I \times k} = 1; \quad QS_{a \times k} = 68 \qquad\qquad\qquad\quad v_{a \times k} = 1$$

Third step of analysis: Each triple of factors is considered. (In this example, there is just one triple combination of factors.)

	$I \times a \times k$							
$I_1 a_1 k_1$	$I_1 a_1 k_2$	$I_1 a_2 k_1$	$I_1 a_2 k_2$	$I_2 a_1 k_1$	$I_2 a_1 k_2$	$I_2 a_2 k_1$	$I_2 a_2 k_2$
-42	-33	$+27$	-13	-33	-32	$+46$	$+30$

$$T^2/N = 39 \qquad N/N_c = 64/8 = 8$$

We calculate:

$$QS_{\alpha \times \beta \times \gamma} = \frac{\text{sum of squares}}{N/N_c} - \frac{T^2}{N} - QS_{\alpha \times \beta} - QS_{\alpha \times \gamma} - QS_{\beta \times \gamma}$$

with

$$v_{\alpha \times \beta \times \gamma} = v_\alpha . v_\beta . v_\gamma$$

degrees of freedom:

$$QS_{I \times a \times k} = 16, \qquad v_{I \times a \times k} = 1. \qquad (\alpha = I, \ \beta = a, \ \gamma = k)$$

Fourth step of analysis: Because there are no more than three different types of factor, we now can consider the total of all values as a single sample:

$$QS_{all} = \text{sum of squares of all values minus } T^2/N$$
$$= 8^2 + 8^2 + 7^2 + \ldots + 7^2 + 7^2 + 8^2 - 39 = 1673$$

with

$$v = N - 1 = 64 - 1 = 63$$

degrees of freedom.

$QS_{rest} = QS_{all}$ minus sum of all other QS-values $= 602$
with

$$v_{rest} = v_{all} \text{ minus sum of all other degrees of freedom}$$
$$= 63 - 7 = 56$$

degrees of freedom.

Final step. We now compare the variance of each factor and each combination of factors with the residual variance. The latter may be interpreted in this example as corresponding to the variance of porosity before diagenesis. If the variance ratio is not significant according to Snedecor's F, then diagenetic factors did not change appreciably the initial porosity distribution:

	Source	QS	v	var.	v.r.	significance
	I	81	1	81	7.4	s.
Factors	a	827	1	827	75.2	s.
	k	33	1	33	3	n.s.
Pair interaction	$I \times a$	42	1	42	3.8	n.s.
	$I \times k$	4	1	4	2.8	n.s.
	$a \times k$	68	1	68	6.2	p.s.
Triple interaction	$I \times a \times k$	16	1	16	1.5	n.s.
Rest effects	others	602	56	11	—	—

v=degrees of freedom   v.r.=variance ratio   n=not,   s=significant,   p=probably
(*Example*: Source: I; v.r.=81/11=7.4)

The results are:

1. Distance and pore contents are significant,
2. Grain size is not significant,
3. There are no interactions except possibly for $a \times k$.

Thus, in view of the last item, the investigation should be repeated with a carefully collected larger sample which might be more representative. Let us assume that the $a \times k$ interaction turns out to be significant. In this situation a and k depend on each other and we cannot consider a and k independently as we did in the first step of our investigation. Hence, the first step reduces to the analysis of pore contents (I) with the already familiar result QS_I=81. As the next step, the set of data may be broken into an a_1-group and an a_2-group each of which is investigated separately. Alternatively, the data may be split into a k_1-group and a k_2-group. If, however, it would have turned out at the initial analysis that triple interaction or each pair interaction is significant, then the whole analysis would break down. For instance, for a significant triple interaction, we could not study the affects of pair interactions and single factors with the methods presented here.

CHAPTER 23

Performance of Statistical Investigations. The Most Important Flaws Encountered in Papers Using Statistical Methods

Performance of statistical investigations

(0) Select the fundamental approach

In many instances the earth scientist wants to describe a material with different properties. Then there is just one approach by using some simple or sophisticated statistical methods. In other instances, however, the scientist wants to simulate geological events and related phenomena by a model, that is, by a set of statements or formulae which are a simplified copy of reality. In this situation there are three possibilities:

(i) The approach must be probabilistic, that is stochastic, and the reader has to use either classical probability theory or Monte Carlo methods and possibly other modern tools such as decision theory.

(ii) The approach must be deterministic.

(iii) The approach can be stochastic as well as deterministic. Then it might be difficult to decide which approach is superior and should be used. The deterministic model results in causal explanations. If the deterministic model is correct within specified limits, then it is superior to any stochastic model: it explains the past and may permit forecasting the future. The stochastic model yields correlations rather than causes. These correlations may be very useful for forecasting but lack principally an insight into the investigated mechanisms. The question "Why is this so?" is not answered.

(1) Planning a statistical investigation

(i) Select the properties to be investigated.

(ii) Estimate which factors are most important and which factors may be relevant.

(iii) Select the geographical locations for sampling if the number of useful locations is abundant.

(iv) Specify the sample sizes if the number of items to be investigated is large.

(v) Select the statistical apparatus to be used.

(vi) Perform preliminary examinations if the situation is blurred or too complex.

The statistical approach must be planned before a series of experiments or observations or other investigations are started, otherwise a set of data may result which is large but worthless. This is a trivial but may be an overlooked consequence. The famous statistician R. A. Fisher once remarked: "To consult the statistician after an experiment is finished is often merely to ask him to conduct a post-mortem examination. He can perhaps say what the experiment died of."

When performing statistical investigations on the elementary level presented in this book, the additional steps usually are

(2) *Draw significant samples*

(3) *Determine the frequency distributions*

(4) *Compute the parameters of the frequency distributions, regression formulae, and correlations*

(5) *Test the significance of the calculated parameters*

(6) *Check the hypothesis that the samples belong to a specified population*

The quantification of the earth sciences by stochastic and deterministic methods seems to be a slow and painful process. It therefore is not surprising that many papers in the realm of geology and paleontology are unsatisfactory from the point of view of the statistician.

Typical flaws of nonstatistical papers in geology and paleontology

(1) A too large set of data is represented. *Example*: Publication of much well data from a rather monotonous region. The publication of one standard profile with a table displaying regional trends would be just as informative.

(2) A nonrepresentative, misleading set of data is published. *Example*: One geological profile from each of ten quarries is published. Within each quarry, the variance of thicknesses is larger than the variance between quarries. The reader, however, gets the impression that each profile is representative for a quarry and that there are systematic changes from quarry to quarry.

(3) Conclusions are drawn from a sample which is too small. *Example*: The dip angle of a rock which covers a large area is estimated from a few measurements at a single location.

(4) Measurements are too accurate. *Example*: The variability of some length is about 10 inches. Then the measurement should not be more accurate than say 1 inch or $\frac{1}{2}$ inch.

(5) Measurements are too inaccurate. In this situation the errors do not cancel when calculating the mean. Further, systematic errors of measurements may become large if one does not work carefully. Last, not least, the variance of inaccurate measurements reflects the generosity of the worker rather than the variance of the material.

(6) The measured properties are not defined properly. This is a rather typical flaw in paleontology when defining distances between certain points of bones and fossils of complex shape.

(7) Statements are based on measurements of other scientists of unknown reliability. Unfortunately, this problem can not be avoided. Nevertheless, it is surprising that different scientists may arrive at significantly different results when measuring the same material.

Typical flaws of statistical papers in geology and paleontology

(1) The data are incomplete or not evaluated fully. As a consequence, the reader of the paper cannot judge the significance of the stated results.

(2) The author invents new statistical parameters of doubtful value, whose significance cannot be tested.

(3) The author gives a rather elaborate explanation of the statistical methods used; such explanations, if really necessary, should be in the appendix.

(4) The text is organized badly. Illuminating remarks that would clarify the reasoning of the author are omitted. Instead of this, the reader is confronted with complex mathematical unfamiliar formulae. There may even be mathematical discussions which are not required to solve the geological problem.

(5) The reader gets the impression that the statistics are an end in itself.

Some Textbooks and Treatises

A. Not much knowledge of mathematics required

MORONEY, M. J. *Facts from Figures*. Penguin Books Ltd., 1956.
SIMPSON, G. G., ROE, A. & LEWONTIN, R. C. *Quantitative Zoology*. Harcourt, Brace & Co., New York—Burlingame, 1960.
WONNACOTT, TH. H. & WONNACOTT, R. J. *Introductory Statistics*. 2nd ed. Wiley, New York, 1977.

B. Little knowledge of mathematics required

FISHER, R. A. *Statistical Methods for Research Workers*. 2nd ed. Oliver & Boyd, London, 1932.
SNEDECOR, G. W. *Statistical Methods*. Iowa State College Press, Ames, Iowa, 1948.
SPIEGEL, M. R. *Theory and Problems of Statistics*. Schaum's Outline Series, Schaum Publ. Co., New York, 1961.

C. Special topics. Not much mathematics required

COOLEY, W. W. & LOHNES, P. R. *Multivariate Data Analysis*. Wiley, New York, 1971.
DRAPER, N. R. & SMITH, H. *Applied Regression Analysis*. Wiley, New York, 1966.
GUENTHER, W. C. *Analysis of Variance*. Prentice-Hall Englewood Cliffs, N. J., 1964.
HARMAN, H. H. *Modern Factor Analysis*. 2nd ed. University of Chicago Press, Chicago, 1968
OVERALL, J. E. & KLETT, C. J. *Applied Multivariate Analysis*. McGraw-Hill, New York, 1972.
SIEGEL, S. *Nonparametric Statistics*. McGraw-Hill, New York, 1956.

D. Textbooks of Geostatistics

AGTERBERG, F. P. *Geomathematics*. Elsevier, Amsterdam, 1974.
CHEENEY, R. F. *Statistical methods in Geology*. Allen & Unwin, 1983.
DAVIS, J. C. *Statistics and Data Analysis in Geology*. 2nd ed. Wiley, New York, 1986.
GUILLAUME, A. *Introduction a la géologie quantitative*. Masson, Paris, 1977.
HARBAUGH, J. W. & BONHAM-CARTER, G. F. *Computer Simultation in Geology*. Wiley Interscience, New York, 1970.
KOCH, G. S. & LINK, R. F. *Statistical Analysis of Geological Data*. Wiley, New York, Vol. 1: 1970, Vol. 2: 1971.
KRUMBEIN, W. C. & GRAYBILL, F. A. *An Introduction to Statistical Models in Geology*. McGraw-Hill, New York, 1965.
MERRIAM, D. F. *Computer Fundamentals for Geologists*. COMPUTe Dartmouth College, Hanover, New Hampshire, 1975.
MILLER, R. L. & KAHN, J. S. *Statistical Analysis in the Geological Sciences*. Wiley, New York, 1965.

SMITH, F. G. *Geological Data Processing: Using FORTRAN IV.* Harper & Row, 1968.
TILL, R. *Statistical Methods for the Earth Scientist.* Macmillan, London, 1974.

E. Treatises of special topics in Geostatistics

ANDERSON, T. W. *The Statistical Analysis of Time Series.* Wiley, New York, 1971.
BAILEY, N. I. J. *The Elements of Stochastic Processes with Applications to the Natural Sciences.* Wiley, New York, 1964.
BLACKITH, R. E. & REYMENT, R. A. *Multivariate Morphometrics.* Academic Press, London, 1974.
CHORLEY, R. J. *Spatial Analysis in Geomorphology.* Methuen, London, 1972.
DAVID, M. *Geostatistical Ore Reserve Estimation.* Elsevier, Amsterdam, 1977.
GRIFFITHS, J. C. *Statistical Methods in Sedimentary Petrography.* In MILNER, H. B., *Sedimentary Petrography*, p. 565–609. George Allen & Unwin Ltd., London, 1962.
GUILLAUME, A. *Analyse de variables regionalisées.* Doin Editeurs, Paris, 1977.
HARBAUGH, J. W., DOVETON, J. H. & DAVIS, J. C. *Probability Methods in Oil Exploration.* Wiley, New York, 1977.
HARBAUGH, J. W. & MERRIAM, D. F. *Computer Applications in Stratigraphic Analysis.* Wiley, New York, 1968.
JÖRESKOG, K. G., KLOVAN, J. E. & REYMENT, R. A. *Geological Factor Analysis.* Elsevier, Amsterdam, 1976.
JOURNEL, A. G. & HUIJBREGTS, CH. *Mining Geostatistics.* 2 vols. Academic Press, London, 1979.
LAFFITE, P. (Main author): *Traité d'informatique géologique.* Masson, Paris, 1972.
MATHERON, G. *Les variables régionalisées et leur estimation.* Masson, Paris, 1965.
MEYER, A. *Statistische Verfahren.* Vol. 5. Glückauf, Essen, 1976.
REYMENT, R. A. *Introduction to Quantitative Paleoecology.* Elsevier, Amsterdam, 1971.
REYMENT, R. A. *Biometrical Methods of Well Logging.* Academic Press, London, 1979.
SALAMON, D. G. & LANCASTER, F. H. *Application of Computer Methods in the Mineral Industry.* South Afr. Inst. Min. Met., Johannesburg, 1973.
SCHWARZACHER, W. *Sedimentation Models and Quantitative Stratigraphy.* Elsevier, Amsterdam, 1975.
SNEATH, P. H. A. & SOKAL, R. R. *Numerical taxonomy.* Freeman, San Francisco, 1973.
SOKAL, R. R. & SNEATH, P. H. A. *Principles of Numerical Taxonomy.* Freeman, San Francisco, 1963.

F. Monographs. Edited Collections

BERRY, B. J. L. & MARBLE, D. F. (Ed.) *Spatial Analysis. A Reader in Statistical Geography.* Prentice-Hall, Englewood Cliffs, N. J. 1968.
CHAYES, F. *Ratio Correlation.* Univ. of Chicago, 1971.
CUTBILL, J. L. (Ed.) *Data Processing in Biology and Geology.* System Assoc. Special Vol. 3, Academic Press, 1971.
DAVIS, J. C. & McCULLAGH, M. J. (Ed.) *Display and Analysis of Spatial Data.* Wiley Interscience, New York, 1975.
DE MARSILY, G. & MERRIAM, D. F. (Eds.) *Predictive geology.* Pergamon Press, Oxford, 1982.
GILL, D. & MERRIAM, D. F. (Eds.) *Geomathematical and Petrophysical Studies in Sedimentology.* Pergamon Press, Oxford, 1979.
GUARESCIO, M., DAVID, M. & HUIJBREGTS, CH. (Eds.) *Advanced Geostatistics in the Mining Industry.* Nato Advanced Study Instit. Series C24—Math. and Phys. Sci., D. Reichel Publ. Co. Dordrecht, Boston, 1976.
HANLEY, J. T. & MERRIAM, D. F. (Eds.) *Microcomputer Applications in Geology.* Pergamon Press, Oxford, 1986.
MANIAK, U. *et al.* Internationale Hydrologische Dekade. Theoretische Hydrologie. Heft 1: Stochastische Verfahren. Deutsche Forschungsgemeinschaft, 1971.
McCAMMON, R. (Ed.) *Concepts in Geostatistics.* Springer, Berlin, 1975.

MERRIAM, D. F.: (Ed.) *Computer Applications in the Earth Sciences.* Plenum Press, New York, 1969.

MERRIAM, D. F.: (Ed.) *Geostatistics.* Plenum Press, New York, 1970.

MERRIAM, D. F.: (Ed.) *Mathematical Models of Sedimentary Processes.* Plenum Press, New York, 1972.

MERRIAM, D. F.: (Ed.) *Computer Applications in the Earth Sciences, an update of the 70s.* Plenum Press, New York, 1981.

MERRIAM, D. F.: (Ed.) *Quantitative Techniques for the Analysis of Sediments.* Pergamon Press, Oxford, 1976.

MERRIAM, D. F.: (Ed.) *Recent Advances in Geomathematics.* Pergamon Press, Oxford, 1978.

MERRIAM, D. F.: (Ed.) *Random Processes in Geology.* Springer, Berlin, 1976.

RENDU, J. M. *An Introduction to Geostatistical Methods of Mineral Evaluation.* Monograph Series, South African Inst. of Mining and Metallugy (SAIMM), 1978.

ROMANOVA, M. A. & SARMANOV, O. V. (Eds.) *Topics in Mathematical Geology.* Consultants Bureau, New York, 1970.

SOUTH AFRICAN INSTITUTE OF MINING AND METALLURGY (SAIMM) Symposium on Mathematical Statistics and Computer Applications in Ore Evaluation. Johannesburg, 1966.

VISTELLIUS, A. E. *Studies in Mathematical Geology.* Plenum Press, 1967.

References

Literature related to the examples

ADAM, K. D. Die Teufels- oder Fuchsenlucken bei Eggenburg (NÖ) IV. Die Mammutreste. *Österr. Akad. d. Wiss. Math.-Naturw. Kl. Denkschr.* **112**, 39–60, Wien, 1966.

ADAM, K. D., FIGGE, U. & MARSAL, D. *Individual Statistics of Fossil Mammals: A Monte Carlo Approach* (In press).

ADAM, K. D. & MARSAL, D. *Statistische Absicherung taxonomischer Einheiten am Beispiel dicerorhiner Nashörner aus dem Pleistozän Mitteleuropas* (in press).

ADLER, R., FENCHEL, W. & PILGER, A. *Statistiche Methoden in der Tektonik I und II. Clausthaler Tektonische Hefte 2 und 4*, Pilger, Clausthal-Zellerfeld, 1965.

ALBERS, J. Taxonomie und Entwicklung einiger Arten von *Vaginulina* D'ORB. aus dem Barrême bei Hannover (Foram). *Mitt. Geol. Staatsinst. Hamburg* **21**, 75–112, Hamburg, 1952.

AZÁROFF, L. V. & BUERGER, M. J. *The Powder Method in X-Ray Crystallography*. McGraw-Hill Book Co., New York—Toronto—London, 1958.

BARTH, T. F. W.: *Die Eruptivgesteine*. In BARTH, T. F. W., CORRENS, C. W., ESKOLA, P. *Die Entstehung der Gesteine*. Springer, Berlin, 1939.

BEHRENS, W. Environment Reconstruction for a Part of the Glen Rose Limestone, Central Texas. *Sedimentology* **4**, 65–111, Amsterdam—New York, 1965.

BENNETT, J. G. Broken Coal. *J. Inst. of Fuel* **10**, 22–39, 107–119, London, 1936.

BERTRAM, G. & HARTUNG, J. Ein Folgetest für Vergleichsuntersuchungen zweier Alternativrehen. *Ärztl. Forsch.* **10**, München—Gräfelfing, 1956.

BERTRAM, G. Sequenzanalyse für zwei Alternativfolgen. *Z. Angew. Math. u. Mech.* **40**, 185–189, Dresden—Berlin, 1960.

BETTENSTAEDT, F. Evolutionsvorgänge bei fossilen Foraminiferen. *Mitt. Geol. Staatsinst. Hamburg* **31**, 385–460, Hamburg, 1962.

BRINKMANN, R. Statistisch-Biostratigraphische Untersuchungen an mitteljurassischen Ammoniten über Artbegriff und Stammesentwicklung. *Abh. Ges. Wiss. Göttingen Math.—phys. Kl. N. F.* **13**, H, 3, 1–249, Weidmann'sche Buchhdlg., Berlin, 1929.

BRUNNACKER, K. Die Lockerbraunerde im Bayerischen Wald. *Geol. Blätter f. Nordost-Bayern* **15**, 65–76, Erlangen, 1965.

BRYSON, R. A. & DUTTON, J. A. Some Aspects of the Variance Spectra of Tree Rings and Varves. *New York Acad. Sci. Ann.* **95**, 580–604, New York, 1961.

CADIGAN, R. A. A Method for Determining the Randomness of Regionally Distributed Quantitative Geologic Data. *J. Sed. Pet.* **32**, 813–818, 1962.

CARSS, B. W. & NEIDELL, N. S. A Geological Cyclicity Detected by Means of Polarity Coincidence Correlation. *Nature* **212**, 136–137, London—New York, 1966.

CHAYES, F. Effect of Change of Origin on Mean and Variance of Twodimensional Fabrics. *Am. J. Sci.* **252**, 567–570, New Haven, 1954.

CHAYES, F. *Petrographic Modal Analysis*. Wiley, New York, 1956.

COCHRAN, W. G. *Design and Analysis of Sampling*. In SNEDECOR, G. W. *Statistical Methods Applied to Experiments in Agriculture and Biology*. Iowa State University Press, Ames, Iowa, 5th ed. 1966.

COLE, L. C. Some Features of Random Population Cycles. *J. Wildlife Management* **18**, 2–24, Menasha, 1954.

CORRENS, C. W. *Die Sedimentgesteine.* In BARTH, T. F. W., CORRENS, C. W., ESKOLA, P. *Die Entstehung der Gesteine.* Springer, Berlin, 1939.

CURRAY, J. R. The Analysis of Two-dimensional Orientation Data. *J. Geology* **64**, 117–131, 1956.

CURRAY, J. R. & GRIFFITHS, J. C. Sphericity and Roundness of Quartz Grains in Sediments. *Bull. Geol. Soc. Am.* **66**, 1075–1096, Washington, 1955.

DAVIS, M. B. *A Method for Determination of Absolute Pollen Frequency.* In KUMMEL, B. & RAUB, D. *Handbook of Paleontological Techniques* p. 674–686. W. H. Freeman Co., San Francisco—London, 1965.

ELTGEN, H. Quantitative biofazielle Untersuchungen an Kalken. (Diss.). Bergakad. Clausthal, 1966.

v. ENGELHARDT, W. *Der Porenraum der Sedimente.* Springer, Berlin—Göttingen—Heidelberg, 1960.

EPSTEIN, B. The Mathematical Description of Certain Breakage Mechanisms Leading to the Logarithmico-Normal Distribution. *J. Franklin Inst.* **244**, 471, Philadelphia, 1947.

EPSTEIN, B. Statistical Aspects of Fracture Problems. *J. appl. Phys.* **19**, 140–147, New York, 1948.

EPSTEIN, B. Application of the Theory of Extreme Values in Fracture Problems. *J. Am. Statist. Assoc.* **43**, 403–412, Washington, 1948.

ESKOLA, P. *Die metamorphen Gesteine.* In BARTH, T. F. W., CORRENS, C. W., ESKOLA, P. *Die Entstehung der Gesteine.* Springer, Berlin, 1939.

FISCHER, A. G. Latitudinal Variations in Organic Diversity. *Evolution* **14**, 64–81, USA, 1960.

FOLK, R. L. A Review of Grain-Size Parameters. *Sedimentology* **6**, 73–93, Amsterdam—New York, 1966.

FÜCHTBAUER, H. Sedimentpetrographische Untersuchungen in der älteren Molasse nördlich der Alpen. *Ecl. geol. Helvet.* **57**, 157–298, Basel, 1964.

FÜCHTBAUER, H. Influence of Different Types of Diagenesis on Sandstone Porosity. 7th World Petrol. Congr., Mexico, Panel Discussion No. 3, 1967.

FÜCHTBAUER, H. *Wechselfolgen.* In v. ENGELHARDT, W., FÜCHTBAUER, H. & MÜLLER, G. *Die Sedimentgesteine.* Schweizerbart, Stuttgart 1968.

FÜCHTBAUER, H. *Sandsteine, Konglomerate und Breccien.* In v. ENGELHARDT, W., FÜCHTBAUER, H. & MÜLLER, G. *Die Sedimentgesteine.* Schweizerbart, Stuttgart, 1968.

GEBELEIN, H. Beiträge zum Problem der Kornverteilungen. *Chemie-Ing.-Techn.* **28**. 773–782, Weinheim/Bergstr., 1956.

GLAGOLEV, A. A. Quantitative Analysis with the Microscope by the Point-Method *Engl Min. J.* **135**, 339–340, 1934.

GORDON, P. *Théorie des chaines de markov finies et ses applications.* Dunod, Paris, 1965.

GRABERT, B. Phylogenetische Untersuchungen an Gaudryina und Spiroplectinata (Foram.) besonders aus dem norddeutschen Apt und Alb. *Abh. senckenberg. naturforsch. Ges.* **498**, 1–71, Frankfurt a. M. 1959.

GRIFFITHS, J. C. Size Versus Sorting in some Caribbean Sediments. *J. Geology* **59**, 211–243, 1951.

GRIFFITHS, J. C. Estimation of Error in Grain Size Analysis. *J. Sed. Pet.* **23**, 75–84, 1953.

GRIFFITHS, J. C. Future Trends in Geomathematics. *Mineral Industries* **35**, No. 5, Pennsylvania State Univ. 1966.

GRIFFITHS, J. C. *Statistical Methods in Sedimentary Petrography.* In H. B. MILNER, *Sedimentary Petrography.* 4th rev. ed. George Allen & Unwin Ltd., London, 1962.

GRIFFITHS, J. C. & ROSENFELD, M. A. Operator Variation in Experimental Research. *J. Geology.* **62**, 74–91, 1954.

HARRIS, C. C. Monte Carlo Model of a Fracture Process. *Nature* **209**, 1302–1303, London—New York, 1966.

INMAN, D. L. Sorting of Sediments in the Light of Fluid Mechanics *J. Sed. Pet.* **19**, 51–70, 1949.

INMAN, D. L. Measures for Describing the Size Distribution of Sediments. *J. Sed. Pet.* **22**, 125–145, 1952.

KAUFMANN, R. Variationsstatistische Untersuchungen über die "Artabwandlung" und "Artumbildung" an der oberkambrischen Trilobitengattung Olenus Dalm. *Abh. Geol. Paläont. Inst. Greifswald* **10**, 1933.

KAUFMANN, R. Exakt nachgewiesene Stammesgeschichte. *Naturwiss.* **22**, 803–807, Berlin, 1934.

KITTLEMAN, JR. L. R. Application of ROSIN's Distribution in Size-Frequency Analysis of Clastic Rocks. *J. Sed. Pet.* **34**, 483–502, 1964.

KOLMOGOROFF, A. N. Über das logarithmisch normale Verteilungsgesetz der Teilchen bei Zerstückelung. *C. R. Acad. Sci. U.R.S.S.* **31**, 99, Leningrad, 1941.

KOLMOGOROFF, A. N. *Solution of a Problem in Probability Theory, Connected with the Problem of the Mechanism of Stratification.* Am. Math. Soc. Transl. No. **53**, 1–8, Providence, 1951.

KOTTLER, F. The Distribution of Particle Sizes. *J. Franklin Inst.* **250**, 339–356, 419–441, Philadelphia, 1950.

KRUMBEIN, W. C. Preferred Orientation of Pebbles in Sedimentary Deposits. *J. Geology* **47**, 673–706, 1939.

KRUMBEIN, W. C. Applications of Statistical Methods to Sedimentary Rocks. *J. Am. Stat. Assoc.* **49**, 51–66, Washington, 1954.

KRUMBEIN, W. C. The "Geological Population" as a Framework for Analysing Numerical Data in Geology, Part 3. *Geol. J.* **2**, 341–368, Liverpool—Manchester, 1960.

KRUMBEIN, W. C. Open and Closed Number System and Stratigraphic Mapping. *Bull. Am. Ass. Petrol. Geol.* **46**, 2229–2245, Tulsa, 1962.

KRUMBEIN, W. C. & MILLER, R. L. Design of Experiments for Statistical Analysis of Geological Data. *J. Geology* **61**, 510–532, 1953.

KRUMBEIN, W. C. & MONK, G. D. Permeability as a Function of the Size Parameters of Unconsolidated Sand. *Petroleum Development and Technology A.I.M.E.* **146** and **151**, 359–369, Philadelphia, 1942/43.

KRUMBEIN, W. C. & SLOSS, L. L. *Stratigraphy and Sedimentation.* 2nd ed. Freeman Co., San Francisco—London, 1963.

KRUMBEIN, W. C. & GRAYBILL, F. A. *An Introduction to Statistical Models in Geology.* McGraw-Hill, New York, 1965.

LEOPOLD, L. B., WOLMAN, M. G. & MILLER, J. P. *Fluvial Processes in Geomorphology.* Freeman Co., San Francisco—London, 1964.

LOTZE, F. Beitrag zue Kenntnis der Mutationen von *Calceola sandalina* (L.). *Senckenbergiana* **10**, 158–169, Frankfurt a. M. 1928.

LÜTTIG, G. Methodische Fragen der Geschiebeforschung. *Geol. Jb.* **75**, 361–418, Hannover, 1958.

MARSAL, D. Zur Methodik der Paläontologie. Die statistische Sicherung von Mittelwerten und Korrelationsziffern. *N. Jb. Min. etc. Mhe. Abt. B, H.* **8**, 248–256, Stuttgart, 1949.

MATUSZAK, D. R. Stratigraphic Trend Studies by Electronic Computer. *World Oil* **163**, 61–65, 92, Houston, 1966.

McCAMMON, R. B. Efficiencies of Percentile Measures for Describing the Mean Size and Sorting of Sedimentary Particles. *J. Geology* **70**, 453–465, 1962.

McCAMMON, R. B. *J. Geology* **74**, No. 5, Part 2, 1966.

MELAND, N., FERM, J. C. & NORRMAN, J. O. On the Problem of n in Weight Frequency Distributions. *J. Sed. Pet.* **35**, 984–985, 1965.

MERRIAM, D. F. Symposium on Cyclic Sedimentation. *Bull.* **169**, Vol. II, Kansas Geological Survey, 1964.

MIDDLETON, G. Y. The Tukey Chi-Square Test. *J. Geology* **73**, 547–549, 1965.

MILLER, R. L. Trend Surfaces: Their Application to Analysis and Description of Environments of Sedimentation. *J. Geology* **64**, 425–446, 1956.

MILLER, R. L. & KAHN, J. ST. *Statistical Analysis in the Geological Sciences*, Wiley, New York—London—Sydney, 1965.

MORONEY, M. J. *Facts from Figures.* Penguin Books Ltd., Harmondsworth, Middlesex, 1956.

MOSIMANN, J. E. *Statistical Methods for the Pollen Analyst: Multinomial and Negative Multinomial Techniques.* In KUMMEL, B. & RAUP, D. *Handbook of Paleontological Techniques* pp. 636–673. Freeman Co., San Francisco—London, 1965.

MÜLLER G. *Sediment-Petrologie Teil I: Methoden der Sedimentuntersuchung.* Schweizerbart, Stuttgart 1964.

MÜLLER G. & FÖRSTNER, U. Sedimenttransport im Mündungsgebiet des Alpenrheins. *Geol. Rdsch.* **58**, 229–259, Berlin, 1958.

NEDERLOF, M. H. Structure and Sedimentology of the Upper Carboniferous of the Upper Pisuerga Valleys, Cantabrian Mountains, Spain. *Leidse Geol. Mededel.*, Deel 24, Aufl. 2, 603–703, Leiden, 1959.

NIGGLI, P. *Gesteine und Minerallagerstätten, 1. Band: Allgemeine Lehre von den Gesteinen und Minerallagerstätten.* Birkhäuser, Basel, 1948.

OELSNER, O. *Grundlagen zur Untersuchung und Bewertung von Erzlagerstätten.* Thüringen-Verlag P. E. Blank, Gera, 1952.

PAPOULIS, A. *Probability, Random Variables and Stochastic Processes.* McGraw-Hill, New York, 1965.

PEARN, W. C. Finding the Ideal Cyclothem. *Kansas Geol. Survey Bull.* **169**, 400–413, 1964.

PETTIJOHN, F. J., POTTER, P. E. & SIEVER, R. *Geology of Sand and Sandstone.* Indiana University Printing Plant, 1965.

PHILIPP, W. *et al.* Zur Geschichte der Migration im Gifhorner Trog. *Erdöl und Kohle, Erdgas, Petrochemi* **16**, 456–468, Hamburg, 1963.

PINCUS, H. J. Some Vector and Arithmetic Operations on Two-Dimensional Orientation Variates, with Applications to Geological Data. *J. Geology* **64**, 533–557, 1956.

PUTNAM, *et al. Trans. Am. Geophys. Union* **30**, 337, New York 1949.

RAUP, O. B. & MIESCH, A. T. A New Method for Obtaining Significant Average Directional Measurements in Cross-Stratification Studies. *J. Sed. Pet.* **27**, 313–321, 1957.

RENSCH, B. *Neuere Probleme der Abstammungslehre.* Ferdinand Enke: Stuttgart, 1947.

RUSNAK, G. A. A Fabric and Petrologic Study of the Pleasantview Sandstone. *J. Sed. Pet.* **27**, 41–55, 1957.

SASIENI, M., YASPAN, A. & FRIEDMAN, L. *Operations Research, Methods and Problems.* Wiley, New York—London, 1959.

SCHACHTSCHABEL, P. & RENGER, M. Beziehung zwischen V- und pH-Wert von Böden. *Z. Pflanzenernährung, Düngung, Bodenkunde* **112**, 238–248, Weinheim/Bergstr., 1966.

SCHEIDEGGER, A. E. *Theoretical Geomorphology.* Springer, Berlin—Göttingen—Heidelberg, 1961.

SCHEIDEGGER, A. E. *On the Statistics of the Orientation of Bedding Planes, Grain Axes, and Similar Sedimentological Data.* U. S. Geol. Survey Prof. Paper 525–C, C164–C167, Washington, 1965.

SCHMID, CH. Ringversuche zur Überprüfung der Zuverlässigkeit von Porositäts- und Permeabilitätsmessungen an Gesteinsproben. *Erdöl u. Kohle* **8**, 442–446, Hamburg, 1953.

SEIBOLD, E. Geological Investigation of Near-Shore Sand-Transport—Examples of Methods and Problems from the Baltic and North Seas. *Progress in Oceanography* **1**, 1–70, Pergamon Press, London, 1963.

SHARP, W. E. & POW-FOONG FAN. A Sorting Index. *J. Geology* **71**, 76, 1963.

SHREIDER, YO, A. *Method of Statistical Testing. Monte Carlo Method.* Elsevier, Amsterdam—London—New York, 1964.

SIMPSON, G. G., ROE, A. & LEWONTIN, R. C. *Quantitative Zoology.* Harcourt, Brace & Co., New York—Burlingame, 1960.

SNEDECOR, G. W. *Statistical Methods.* 4th ed. The Iowa State College Press, Ames, Iowa, 1948.

SOERGEL, W. *Die Jagd der Vorzeit* Gustav Fischer, Jena, 1922.

STEINMETZ, R. Analysis of Vectorial Data. *J. Sed. Pet.* 801–812, 1962.

SYLVESTER-BRADLEY, P. C. The Description of Fossil Populations. *J. Paleont.* **32**, 214–235, Tulsa, 1958.

TAKÁCS, L. *Processus Stochastiques. Problèmes et Solutions.* Dunod, Paris, 1964.

TALASH, A. W. & CRAWFORD, P. B. Rock Properties Computed from Random Pore Size Distribution. *J. Sed. Pet.* **35**, 917–921, 1965.

TEUSCHER, E. O. Methodisches zur quantitativen Strukturgliederung körniger Gesteine. *Min. Petr. Mitt.* **44**, 410–421, Wien, 1933.

TEUSCHER, E. O. Quantitative Kennzeichnung der westerzgebirgischen Granite. *N. Jb. Min. etc.* Beil.-Bd. 69, Abt. A, 415–459, Stuttgart, 1935.

THEIMER, O. Über die "statistische Mechanik" von Zermahlungsvorgängen. *Kolloid-Zeitschrift* 128, 1–6, Dresden, 1952.

TUNN, W. Ermittlung einiger wichtiger gesteinsphysikalischer Daten von Erdöl- und Erdgas-Speichergesteinen und Verglich der mit verschiedenen Methoden gewonnenen Werte. *Erdöl-Erdgas-Zeitschrift* 82, 404–413, Hamburg—Wien, 1966.

VISTELIUS, A. B. Über die Frage des Bildungsmechanismus von Schichten. *Akad. Nauk SSR, Doklady* 65, 191–194, Leningrad, 1949.

VISTELIUS, A. B. Probleme der mathematischen Geologie. *Z. Angew. Geol.* 11, 265–268, 306–313, 356–359, Berlin, 1965.

WALGER, E. Zur Darstellung von Korngrößenverteilungen. *Geol. Rdsch.* 54, 976–1002, Berlin, 1965.

WANLESS, H. R. *Pennsylvanian Correlations in the Eastern Interior and Appalachian Coal Fields.* Geol Soc. Am. Special Papers No. 17, Washington, 1939.

WEIGELT, J. Die tektonische Prädestination des Lebensraumes alttertiärer Wirbeltierfaunen Deutschlands. *Z. dt. geol. Ges.* 100, 410–426, Stuttgart, 1950.

WELTE, D. H. Kohlenwasserstoffgenese in Sedimentgesteinen: Untersuchungen über den thermischen Abbau von Kerogen unter besonderer Berücksichtigung der n-Paraffinbildunge. *Geol. Rdsch.* 55, 131–144, Berlin, 1966.

WENGER, R. Die germanischen Ceratiten. *Palaeontographica Abt. A*, 108, Lfg. 1–4, 57–129, Stuttgart, 1957.

WILKS, S. S. *Mathematical Statistics.* Princeton University Press, Princeton, N. J., 1946.

Tables

Table I: Power Sums

$$S_k = 2(1^k + 2^k + 3^k + \ldots + m^k) \qquad\qquad m = \tfrac{1}{2}(n-1),\ n\ odd$$
$$S_k = 2(1^k + 3^k + 5^k + \ldots + (n-1)^k) \qquad\qquad\qquad neven$$
$$k = 2, 4, 6, \ldots$$

Table II: The Gaussian or Normal Distribution

Calculate $z = (x - \bar{x})/\sigma$. Omit the sign if z is negative. Read the corresponding values of σY and F^*. Y is the ordinate of the Gaussian curve. F^* is the cumulative relative frequency in percentage between $z = 0$ and $z = (x - \bar{x})/\sigma$. The relative cumulative frequency between $x = z = -\infty$ and $z = (x - \bar{x})/\sigma$ is $F = 50 + F^*$ if x is larger than $\bar{x}$. It is equal to $F = 50 - F^*$ if x is smaller than $\bar{x}$.

Examples: $\bar{x} = 3$, $\sigma = 2$:

$x = 5$: $z = (5-3)/2 = 1$ $2Y = 0.242$ $Y = 0.121$ $F^* = 34.13\%$ $F\% = 84.13\%$
$x = 1$: $z = (1-3)/2 = -1$ $2Y = 0.242$ $Y = 0.121$ $F^* = 34.13\%$ $F\% = 15.87\%$

Table III: The Student-t Distribution

Example: Degrees of freedom: $v = 30$: $t = 2.75$ at the confidence level 99% (i.e., the chance is 1% that the considered statistic parameter is outside of the calculated confidence interval).

Table IV: Snedecor's F-table

Example: Let the smaller value of $\hat{s}^2$ have 2 degrees of freedom, and let the larger value of $\hat{s}^2$ have 3 degrees of freedom. Then $F = 19.2$ if the chance is 5% that the test fails.

Table V: Chi Square

Example: For $v = 2$ degrees of freedom, the chance is 70% of not exceeding the Chi-square value 2.4.

Table VI: Testing the Significance of the Coefficient of Correlation

Example: The sample size is $N=3$; that is the sample consists of three pairs of numbers. Linear correlation is justified with a chance of 95% if r is larger than 0.997 or smaller than -0.997 (the largest possible value of r is $+1$, the smallest value is -1).

Table VII: Random Numbers

TABLE I. *Power sums*

n	S_2	S_4	S_6
2	2	2	2
4	20	164	1 460
6	70	1 414	32 710
8	168	6 216	268 008
10	330	19 338	1 330 890
12	572	48 620	4 874 012
14	910	105 742	14 527 630
16	1 360	206 992	37 308 880
18	1 938	374 034	85 584 018
20	2 660	634 676	179 675 780
22	3 542	1 023 638	351 208 022
24	4 600	1 583 320	647 279 800
26	5 850	2 364 570	1 135 561 050
28	7 308	3 427 452	1 910 402 028
30	8 990	4 842 014	3 100 048 670

n	S_2	S_4	S_6
3	2	2	2
5	10	34	130
7	28	196	1 588
9	60	708	9 780
11	110	1 958	41 030
13	182	4 550	134 342
15	280	9 352	369 640
17	408	17 544	893 928
19	570	30 666	1 956 810
21	770	50 666	3 956 810
23	1 012	79 948	7 499 932
25	1 300	121 420	13 471 900
27	1 638	178 542	23 125 518
29	2 030	255 374	38 184 590
31	2 480	356 624	60 965 840
33	2 992	487 696	94 520 272
35	3 570	654 738	142 795 410
37	4 218	864 690	210 819 858
39	4 940	1 125 332	304 911 620
41	5 740	1 445 332	432 911 620
43	6 622	1 834 294	604 443 862
45	7 590	2 302 806	831 203 670
47	8 648	2 862 488	1 127 275 448
49	9 800	3 526 040	1 509 481 400
51	11 050	4 307 290	1 997 762 650
53	12 402	5 221 242	2 615 594 202
55	13 860	6 284 124	3 390 435 180
57	15 428	7 513 436	4 354 215 788
59	17 110	8 927 998	5 543 862 430
61	18 910	10 547 998	7 001 862 430

TABLE II. *The Normal Distribution*

z	σ.Y	F*	z	σ.Y	F*	z	σ.Y	F*
0.00	0.3989	0%	1.35	0.1604	41.15	2.70	0.0104	49.65
0.05	0.3984	2.00	1.40	0.1497	41.92	2.75	0.0091	49.70
0.10	0.3970	3.98	1.45	0.1394	42.65	2.80	0.0079	49.74
0.15	0.3945	5.96	1.50	0.1295	43.32	2.85	0.0069	49.78
0.20	0.3910	7.93	1.55	0.1200	43.94	2.90	0.0060	49.81
0.25	0.3867	9.87	1.60	0.1109	44.52	2.95	0.0051	49.84
0.30	0.3814	11.79	1.65	0.1023	45.05	3.00	0.0044	49.87
0.35	0.3752	13.68	1.70	0.0941	45.54	3.05	0.0038	49.89
0.40	0.3683	15.54	1.75	0.0863	45.99	3.10	0.0033	49.90
0.45	0.3605	17.36	1.80	0.0790	46.41	3.15	0.0028	49.92
0.50	0.3521	19.15	1.85	0.0721	46.78	3.20	0.0024	49.93
0.55	0.3429	20.88	1.90	0.0656	47.13	3.25	0.0020	49.94
0.60	0.3332	22.58	1.95	0.0596	47.44	3.30	0.0017	49.95
0.65	0.3230	24.22	2.00	0.0540	47.73	3.35	0.0015	49.96
0.70	0.3123	25.80	2.05	0.0488	47.98	3.40	0.0012	49.97
0.75	0.3011	27.34	2.10	0.0440	48.21	3.45	0.0010	49.97
0.80	0.2897	28.81	2.15	0.0396	48.42	3.50	0.0009	49.98
0.85	0.2780	30.23	2.20	0.0355	48.61	3.55	0.0008	49.98
0.90	0.2661	31.59	2.25	0.0317	48.78	3.60	0.0006	49.98
0.95	0.2541	32.89	2.30	0.0283	48.93	3.65	0.0005	49.99
1.00	0.2420	34.13	2.35	0.0252	49.06	3.70	0.0004	49.99
1.05	0.2300	35.31	2.40	0.0224	49.18	3.75	0.0004	49.99
1.10	0.2179	36.43	2.45	0.0198	49.29	3.80	0.0003	49.99
1.15	0.2059	37.49	2.50	0.0175	49.38	3.85	0.0002	49.99
1.20	0.1942	38.49	2.55	0.0155	49.46	3.90	0.0002	50.00
1.25	0.1827	39.44	2.60	0.136	49.53	3.95	0.0002	50.00
1.30	0.1714	40.32	2.65	0.119	49.60	4.00	0.0001	50.00

TABLE III. *Student-t Distribution*

v	$t_{95\%}$	$t_{99\%}$	v	$t_{95\%}$	$t_{99\%}$	v	$t_{95\%}$	$t_{99\%}$
1	12.71	63.66	13	2.16	3.01	30	2.04	2.75
2	4.30	9.92	14	2.14	2.98	40	2.02	2.70
3	3.18	5.84	15	2.13	2.95	50	2.01	2.67
4	2.78	4.60	16	2.12	2.92	60	2.00	2.66
5	2.57	4.03	17	2.11	2.90	70	2.00	2.65
6	2.45	3.71	18	2.10	2.88	80	1.99	2.64
7	2.36	3.50	19	2.09	2.86	90	1.99	2.64
8	2.31	3.36	20	2.09	2.84	100	1.99	2.63
9	2.26	3.25	22	2.07	2.82	150	1.98	2.61
10	2.23	3.17	24	2.06	2.80	200	1.97	2.60
11	2.20	3.11	26	2.06	2.78	300	1.97	2.60
12	2.18	3.06	28	2.05	2.76	∞	1.96	2.58

TABLE IV. *Snedecor's F-Table*
5% Points
Degrees of freedom for greater variance

	1	2	3	4	5	7	10	20	50	∞
1	161	200	216	225	230	237	242	248	252	254
2	18.5	19.0	19.2	19.3	19.3	19.4	19.4	19.4	19.5	19.5
3	10.1	9.6	9.3	9.1	9.0	8.9	8.8	8.7	8.6	8.5
4	7.7	6.9	6.6	6.4	6.3	6.1	6.0	5.8	5.7	5.6
5	6.6	5.8	5.4	5.2	5.1	4.9	4.7	4.6	4.4	4.4
7	5.6	4.7	4.4	4.1	4.0	3.8	3.6	3.4	3.3	3.2
10	5.0	4.1	3.7	3.5	3.3	3.1	3.0	2.8	2.6	2.5
14	4.6	3.7	3.3	3.1	3.0	2.8	2.6	2.4	2.2	2.1
20	4.4	3.5	3.1	2.9	2.7	2.5	2.4	2.1	2.0	1.8
50	4.0	3.2	2.8	2.6	2.4	2.2	2.0	1.8	1.6	1.4
∞	3.8	3.0	2.6	2.4	2.2	2.0	1.8	1.6	1.4	1.0

left column: Degrees of freedom for smaller variance

1% Points
Degrees of freedom for greater variance

	1	2	3	4	5	7	10	20	50	∞
1	4052	4999	5403	5625	5764	5928	6056	6208	6286	6366
2	98	99	99	99	99	99	99	99	99	100
3	34	31	29	29	28	28	27	27	26	26
4	21	18	17	16	16	15	15	14	14	13
5	16	13	12	11	11	10	10	10	9.2	9.0
7	12	9.6	8.5	7.9	7.5	7.0	6.6	6.2	5.9	5.7
10	10	7.6	6.6	6.0	5.6	5.2	4.9	4.4	4.1	3.9
14	8.9	6.5	5.6	5.0	4.7	4.3	3.9	3.5	3.2	3.0
20	8.1	5.9	4.9	4.4	4.1	3.7	3.4	2.9	2.6	2.4
50	7.2	5.1	4.2	3.7	3.4	3.0	2.7	2.3	1.9	1.7
∞	6.6	4.6	3.8	3.3	3.0	2.6	2.3	1.9	1.5	1.0

left column: Degrees of freedom for smaller variance

TABLE V. *Chi Square*

v	$\chi^2_{50\%}$	$\chi^2_{70\%}$	$\chi^2_{80\%}$	$\chi^2_{90\%}$	$\chi^2_{95\%}$	$\chi^2_{99\%}$	$\chi^2_{99.5\%}$	$\chi^2_{99.9\%}$
1	0.5	1.1	1.6	2.7	3.8	6.6	7.9	10.8
2	1.4	2.4	3.2	4.6	6.0	9.2	10.6	13.8
3	2.4	3.7	4.6	6.3	7.8	11.3	12.8	16.3
4	3.4	4.9	6.0	7.8	9.5	13.3	14.9	18.5
5	4.4	6.1	7.3	9.2	11.1	15.1	16.7	20.5
6	5.4	7.2	8.6	10.6	12.6	16.8	18.5	22.5
7	6.4	8.4	9.8	12.0	14.1	18.5	20.3	24.3
8	7.3	9.5	11.0	13.4	15.5	20.1	22.0	26.1
9	8.3	10.7	12.2	14.7	16.9	21.7	23.6	27.9
10	9.3	11.8	13.4	16.0	18.3	23.2	25.2	29.6
11	10.3	12.9	14.6	17.3	19.7	24.7	26.8	31.3
12	11.3	14.0	15.8	18.6	21.0	26.2	28.3	32.9
13	12.3	15.1	17.0	19.8	22.4	27.7	29.8	34.5
14	13.3	16.2	18.2	21.1	23.7	29.1	31.3	36.1
15	14.3	17.3	19.3	22.3	25.0	30.6	32.8	37.7
16	15.3	18.4	20.5	23.5	26.3	32.0	34.3	39.3
17	16.3	19.5	21.6	24.8	27.6	33.4	35.7	40.8
18	17.3	20.6	22.8	26.0	28.9	34.8	37.2	42.3
19	18.3	21.7	23.9	27.2	30.1	36.2	38.6	43.8
20	19.3	22.8	25.0	28.4	31.4	37.6	40.0	45.3
21	20.3	23.9	26.2	29.7	32.7	38.9	41.4	46.8
22	21.3	24.9	27.3	30.8	33.9	40.3	42.8	48.3
23	22.3	26.0	28.4	32.0	35.2	41.6	44.2	49.7
24	23.3	27.1	29.6	33.2	36.4	43.0	45.6	51.2
25	24.3	28.2	30.7	34.4	37.7	44.3	46.9	52.6
26	25.3	29.3	31.8	35.6	38.9	45.6	48.3	54.1
27	26.3	30.3	32.9	36.7	40.1	47.0	49.6	55.5
28	27.3	31.4	34.0	37.8	41.3	48.3	51.0	56.9
29	28.3	32.4	35.1	39.1	42.6	49.6	52.3	58.3
30	29.3	33.5	36.3	40.3	43.8	50.9	53.7	59.7

TABLE VI. *Confidence limits for Correlation Coefficient*

N	$r_{95\%}$	$r_{99\%}$	N	$r_{95\%}$	$r_{99\%}$
3	0.997	1.000	12	0.576	0.708
4	0.950	0.990	14	0.533	0.662
5	0.878	0.959	16	0.498	0.623
6	0.812	0.917	18	0.469	0.590
7	0.754	0.874	20	0.444	0.562
8	0.707	0.835	30	0.360	0.460
9	0.666	0.798	40	0.310	0.402
10	0.633	0.765	120	0.182	0.236

TABLE VII. *Thousand Random Numbers*

39615	80428	26575	30883	92157	47689	05224	99154
63348	72754	55999	10097	42257	76285	98190	86847
30374	12063	17564	62738	44013	52636	18912	18317
04024	18845	16296	77510	19636	70335	57412	10909
98427	49618	62765	53060	98710	42808	34693	30235
83534	25015	40653	51826	11467	97006	70163	32261
44472	66520	18494	05554	22965	17075	27022	14777
42876	18062	18278	11491	82525	12856	46949	18373
57910	41569	29832	43839	18327	53412	66999	89286
57166	00844	96118	88974	29812	74014	55612	55657
72189	21468	50717	19662	98038	99447	21457	80059
02861	53890	73533	15227	36907	88428	08002	11398
13245	97907	00415	08075	54263	08303	89025	53416
21245	64798	33726	97025	39296	03862	95452	34176
80725	97422	86563	35726	23186	21804	59917	94270
90012	94567	46138	18688	69721	23541	17361	35605
61399	68570	72816	45045	50432	83556	37980	59934
17626	76886	42807	91537	94618	81250	75366	46503
18745	81304	55919	97248	92027	58861	35970	91567
97429	98059	33939	49672	94964	66345	19687	57948
00374	15740	84776	36036	81480	11163	73817	95876
20764	13206	34374	20633	03237	13142	30129	56170
47547	74580	14905	58807	40488	49071	95931	40206
63520	22238	53961	36782	05908	39147	02338	37937
36969	60578	93898	41012	33300	64482	43040	25940

The Mathematical Background of Chapter 14

The method presented in Chapter 14 seems to be rather old but practically forgotten. I located it in the small book of G. Schulz, *Formelsammlung zur praktischen Mathematik* (Formulae of practical mathematics), Goeschen, Berlin, 1937. Obviously, the booklet is no longer available. I, therefore, will give a short introduction. Because of its generality, the notation is slightly different from the notation in Chapter 14. Some knowledge of numerical mathematics is required.

Performing the transformation from x to u and applying the least-squares principle of Gauss to the polynomial $y = a_0 + a_1 u + a_2 u^2 + \ldots + a_r u^r$, the following normal equations result which determine the coefficients $a_0, a_1, a_2, \ldots, a_r$:

$$S_{i-1}a_0 + S_i a_1 + S_{i+1}a_2 + \ldots + S_{i-1+r}a_r = [u^{i-1}y] \qquad i = 1,2,3,\ldots,r+1$$

where

$$[u^{i-1}y] = \sum_{j=1}^{n} u_j^{i-1} y_j$$

and

$$S_k = \begin{cases} n \text{ for } k=0 \\ 0 \text{ for } k \text{ odd} \\ 2(1^k + 2^k + 3^k + \ldots + m^k) \text{ for } k \text{ even, } n \text{ odd} \\ 2(1^k + 3^k + 5^k + \ldots + (n-1)^k) \text{ for } k \text{ even, } n \text{ even} \\ \qquad m = \tfrac{1}{2}(n-1) \end{cases}$$

S_k vanishes for all odd values of k. As a consequence, the system of normal equations reduces to two separate systems of linear equations which can be solved separately:

$$y = a_0 + a_1 u + a_2 u^2: \qquad \begin{aligned} na_0 + S_2 a_2 &= [y] \\ S_2 a_0 + S_4 a_2 &= [u^2 y] \end{aligned}$$

$$a_1 = [uy]/S_2$$

171

$$y = a_0 + a_1 u + a_2 u^2 + a_3 u^3:$$

$$na_0 + S_2 a_2 = [y] \qquad S_2 a_1 + S_4 a_3 = [uy]$$
$$S_2 a_0 + S_4 a_2 = [u^2 y] \qquad S_4 a_1 + S_6 a_3 = [u^3 y]$$

$$y = a_0 + a_1 u + a_2 u^2 + a_3 u^2 + a_3 u^3 + a_4 u^4$$

$$na_0 + S_2 a_2 + S_4 a_4 = [y] \qquad S_2 a_1 + S_4 a_3 = [uy]$$
$$S_2 a_0 + S_4 a_2 + S_6 a_4 = [u^2 y] \qquad S_4 a_1 + S_6 a_3 = [u^3 y]$$
$$S_4 a_0 + S_6 a_2 + S_8 a_4 = [u^4 y]$$

and so on.

Remarks to Chapter 15
(Discriminant Analysis)

The mathematical theory of discriminant analysis requires some familiarity with partial differentiation. A particularly simple presentation is in the book of A. Linder, *Statistiche Methoden*, p. 238 ff., Birkhaeuser, Basel, 1960. There also are numerical examples from the realm of paleontology.

To obtain the parameters a and b of the formula $X = ax + by$, solve the equations

$$aS_{11} + bS_{12} = D_x$$
$$aS_{12} + bS_{22} = D_y,$$

for a and b. To obtain the parameters a, b, and c of the formula $X = ax + by + cz$, solve the equations

$$aS_{11} + bS_{12} + cS_{13} = D_x$$
$$aS_{12} + bS_{22} + cS_{23} = D_y$$
$$aS_{13} + bS_{23} + cS_{33} = D_z.$$

for a, b, and c. The general situation is obvious. The optimal solutions are $X = Ax + By$ and $X = Ax + By + Cz$ respectively where

$$a{:}b = A{:}B \quad \text{and} \quad a{:}b{:}c = A{:}B{:}C$$

respectively. This shows that there is an infinity of equivalent optimal solutions. In Chapter 15 the solutions $X = x + (b/a)y$, that is $A = 1$ and $B = b/a$, and $X = x + (b/a)y + (c/a)z$, that is $A = 1$, $B = b/a$, and $C = c/a$, are used.

The variables x, y, z, ... have in general different units (x may be a length, y a volume, etc.). The dimensions of the parameters a, b, c, ... and A, B, C, ... are such that X is dimensionless. This shows that in discriminant analysis a cocktail of any amount of properties can be mixed.

Some special situations. I restrict the discussion to the formula discussion $X = ax + by$ and the corresponding two equations for a and b (see previous). Let $S_{11}S_{22} - S_{12}^2 \neq 0$. If the determinant

$$\begin{vmatrix} D_x & S_{12} \\ D_y & S_{22} \end{vmatrix}$$

vanishes, then $a=0$ and $X=y$ is the optimum. If the determinant

$$\begin{vmatrix} S_{11} & D_x \\ S_{12} & D_y \end{vmatrix}$$

vanishes, then $b=0$ and $X=x$ is the optimum. If $D_x=D_y$, then $a=b=0$. In all these situations discriminant analysis does not improve the results of statistical analysis of single properties. There are unfavorable situations correspondingly for discriminant analysis with more than two parameters.

Index